The Dopaminergic Mind in Human Evolution and History

W0042094

What does it mean to be human? There are many theories of the evolution of human behavior which seek to explain how our brains evolved to support our unique abilities and personalities. Most of these have focused on the role of brain size or specific genetic adaptations of the brain. In contrast, Fred Previc presents a provocative theory that high levels of dopamine, the most widely studied neurotransmitter, account for all major aspects of modern human behavior. He further emphasizes the role of epigenetic rather than genetic factors in the rise of dopamine. Previc contrasts the great achievements of the dopaminergic mind with the harmful effects of rising dopamine levels in modern societies and concludes with a critical examination of whether the dopaminergic mind that has evolved in humans is still adaptive to the health of humans and to the planet in general.

Fred H. Previc is currently a science teacher at the Eleanor Kolitz Academy in San Antonio, Texas. For over twenty years, he was a researcher at the United States Air Force Research Laboratory where he researched laser bioeffects, spatial disorientation in flight, and various topics in sensory psychology, physiological psychology, and cognitive neuroscience. Dr. Previc has written numerous articles on the origins of brain lateralization, the neuropsychology of 3-D space, the origins of human intelligence, the neurochemical basis of performance in extreme environments, and the neuropsychology of religion.

This book is dedicated to mati and oce (in memoriam) and to Nancy, Andrew and Nicholas.

The Dopaminergic Mind in Human Evolution and History

Fred H. Previc

CAMBRIDGE
UNIVERSITY PRESS

CAMBRIDGE UNIVERSITY PRESS
Cambridge, New York, Melbourne, Madrid, Cape Town,
Singapore, São Paulo, Delhi, Tokyo, Mexico City

Cambridge University Press
The Edinburgh Building, Cambridge CB2 8RU, UK

Published in the United States of America by Cambridge University Press, New York

www.cambridge.org
Information on this title: www.cambridge.org/9780521360890

© Fred H. Previc 2009

This publication is in copyright. Subject to statutory exception
and to the provisions of relevant collective licensing agreements,
no reproduction of any part may take place without the written
permission of Cambridge University Press.

First published 2009
First paperback edition 2011

A catalogue record for this publication is available from the British Library

Library of Congress Cataloguing in Publication data

Previc, Fred H.
    The dopaminergic mind in human evolution and history / Fred H. Previc.
        p. cm.
    Includes bibliographical references and index.
    ISBN 978-0-521-51699-0 (hardback)
    1. Dopaminergic mechanisms. 2. Brain-Evolution. 3. Human evolution.
    4. Neuropsychology. I. Title.
    [DNLM: 1. Receptors, Dopamine–physiology. 2. Behavior–physiology.
    3. Brain Chemistry-genetics 4. Dopamine-physiology.
    5. Evolution. 6. Humans-genetics.
    WL 102.8 P944d 2009]
    QP364.7.p74 2009
    612.8'2–dc22
    2009004657

ISBN 978-0-521-51699-0 Hardback
ISBN 978-0-521-36089-0 Paperback

Cambridge University Press has no responsibility for the persistence or
accuracy of URLs for external or third-party internet websites referred to in
this publication, and does not guarantee that any content on such websites is,
or will remain, accurate or appropriate.

Contents

List of figures   *page* vii
List of tables   viii
Acknowledgments   ix

1 What makes humans special?   1
   1.1 Myths concerning the origins of human behavior   3
      1.1.1 Was human intelligence genetically selected?   3
      1.1.2 Did our larger brains make us more
            intelligent?   10
   1.2 The evolution of human intelligence: an
      alternative view   13
      1.2.1 Dopamine and advanced intelligence   13
      1.2.2 The rise of dopamine during human evolution   17

2 Dopamine in the brain   19
   2.1 The neurochemistry of dopamine   19
   2.2 The neuroanatomy of dopamine   23
   2.3 Dopamine and the left hemisphere   31
   2.4 Dopamine and the autonomic nervous system   33
   2.5 Summary   35

3 Dopamine and behavior   37
   3.1 Dopamine and distant space and time   38
      3.1.1 Dopamine and attention to spatially and temporally
            distant cues   41
      3.1.2 Dopamine and goal-directedness   46
      3.1.3 Dopamine and extrapersonal experiences   49
   3.2 Dopamine and intelligence   53
      3.2.1 Motor programming and sequencing   57
      3.2.2 Working memory   59
      3.2.3 Cognitive flexibility   59
      3.2.4 Abstract representation   61
      3.2.5 Temporal analysis/processing speed   62
      3.2.6 Generativity/creativity   63
   3.3 Dopamine and emotion   64
   3.4 The dopaminergic personality   66
      3.4.1 Ventromedial dopaminergic traits   68

    3.4.2 Lateral-dopaminergic traits   69
    3.4.3 Dopamine and the left-hemispheric (masculine) style   71
  3.5 Summary   73

4 Dopamine and mental health   75
  4.1 The "hyperdopaminergic" syndrome   75
  4.2 Disorders involving primary dopamine dysfunction   79
    4.2.1 Attention-deficit/hyperactivity disorder   79
    4.2.2 Autism   81
    4.2.3 Huntington's disease   83
    4.2.4 Mania (bipolar disorder)   84
    4.2.5 Obsessive-compulsive disorder   86
    4.2.6 Parkinson's disease   88
    4.2.7 Phenylketonuria   90
    4.2.8 Schizophrenia   91
    4.2.9 Tourette's syndrome   95
  4.3 Summary   97

5 Evolution of the dopaminergic mind   101
  5.1 The importance of epigenetic inheritance   101
  5.2 Evolution of the protodopaminergic mind   104
    5.2.1 Environmental adaptations in the "cradle of humanity"   104
    5.2.2 Thermoregulation and its consequences   108
  5.3 The emergence of the dopaminergic mind in later evolution   114
    5.3.1 The importance of shellfish consumption   117
    5.3.2 The role of population pressures and cultural exchange   119
  5.4 Summary   121

6 The dopaminergic mind in history   123
  6.1 The transition to the dopaminergic society   123
  6.2 The role of dopaminergic personalities in human history   130
    6.2.1 Alexander the Great   134
    6.2.2 Christopher Columbus   136
    6.2.3 Isaac Newton   139
    6.2.4 Napoleon Bonaparte   142
    6.2.5 Albert Einstein   144
    6.2.6 Dopaminergic personalities in history – reprise   147
  6.3 The modern hyperdopaminergic society   149
  6.4 Summary   153

7 Relinquishing the dopaminergic imperative   155
  7.1 Reaching the limits of the dopaminergic mind   155
  7.2 Tempering the dopaminergic mind   161
    7.2.1 Altering dopamine with individual behavior   161
    7.2.2 Knocking down the pillars of the hyperdopaminergic society 165
  7.3 Toward a new consciousness   170

References   173
Index   208

Figures

2.1 The chemical structure of dopamine and
norepinephrine. *page* 20

2.2 The dopamine neuron and synapse. 22

2.3 The cardinal directions and nomenclature used in
brain anatomical localization. 24

2.4 Some of the major dopamine systems, shown
in a mid-sagittal view. 26

3.1 The realms of interaction in 3-D space and
their cortical representations. 39

3.2 Upward dopaminergic biases. 43

3.3 The dopaminergic exploration of distant space
across mammals. 44

3.4 An axial (horizontal) section of a human brain
showing reduced dopamine D_2 receptor binding
(increased dopamine activity) in the left and right
caudate nuclei in a reversal shift memory task. 54

5.1 The hypothesized direction of modern human
origins and migration. 116

6.1 Five famous dopaminergic minds in history:
Alexander the Great, Christopher Columbus,
Isaac Newton, Napoleon Bonaparte, and
Albert Einstein. 133

6.2 The progression of the dopaminergic mind. 154

7.1 Restoring balance to the dopaminergic mind. 172

Tables

3.1 Features of the two dopaminergic systems.   *page* 67
4.1 Features of the major hyperdopaminergic disorders.   98
4.2 Co-morbidity of the major hyperdopaminergic disorders.   99
6.1 Dopaminergic traits in famous men of history.   148

Acknowledgments

I wish to thank the many scientists who shared their ideas or findings with me and especially those who reviewed either large sections of this book or the book in its entirety (Dr. Britt Bousman, Mr. Jeff Cooper, Dr. Michael Corballis, Dr. Jaak Panksepp, and Dr. Julie Sherman). I also wish to thank Andrew Peart for his support in making the publication of this work possible.

1   What makes humans special?

Between two and three million years ago, a small creature hardly larger than a pygmy chimpanzee but with a much larger brain relative to its body weight began a remarkable journey. The initial part of that journey didn't involve much by today's standards, merely the ability to scavenge and possibly chase-hunt the creatures of the sub-Saharan African savannahs, to make some rather modest stone-flaked tools for that purpose, and eventually to migrate over the African and possibly the Eurasian land mass. This little creature, arguably our first unequivocally human ancestor, was known as *Homo habilis* ("domestic" man). How the modest abilities of this first human emerged and were transformed into the prodigious human achievements and civilization that exist today is arguably the most important scientific mystery of all. The solution to this mystery will not only help to explain where and why we evolved as we did – it will additionally shed light on how we may continue to evolve in the future.

But, first, some basic questions must be asked, including: what is human nature and what is the basis of it? How much of human nature is related to our genes? Is human nature related to the size and shape or lateralization of our brain? How did human nature evolve? Although our hairless skin and elongated body make our appearance quite different from our primate cousins, it is not our anatomy but our unique brain and behavior that most people consider special. Typical behaviors considered uniquely human include propositional (grammatical) language, mathematics, advanced tool use, art, music, religion, and judging the intent of others. However, outside of religion, which has yet to be documented in any other extant species, at least one other – and, in some cases, several – advanced species have been shown to possess one or more of the above traits. For example, dolphins understand and can use simple grammar in their contact with humans (Herman, 1986) and probably use even more sophisticated grammar in their own ultrasonic communications. Certain avian species such as parrots can count up to ten (Pepperberg, 1990) and, like apes, use mathematical concepts such

1

as similarity and transitivity (Lock and Colombo, 1996). Orangutans display highly advanced tool use, including the preparation of tools for use in procuring food (van Schaik, 2006). As regards music and art, singing is a highly developed and plastic form of communication in songbirds (Prather and Mooney, 2004), apes have proven to be adept musical instrumentalists in their drumming (Fitch, 2006), and elephants and chimpanzees have been known to create realistic and abstract paintings.[1] Finally, chimpanzees (but not monkeys) are able to determine the mental states of others and to engage in mirror self-recognition (Lock and Colombo, 1996), attributes normally considered part of a general mental capability known as the "theory of mind" (see later chapters).

What mostly defines humans, then, is not a unique ability to engage in a particular behavior but rather *the way in which we perform it*. Three features of human behavior are particularly salient: its context-independence, its generativity, and its degree of abstraction. Context-independent cognition, emphasized in the comparative analysis of Lock and Colombo (1996), refers to the ability to perform mental operations on new and different types of information in different settings. The behavior of chimpanzees may be viewed as much more contextually dependent than that of humans because it differs considerably depending on whether they are in the wild or in captivity; in the wild, for example, chimpanzees are relatively more likely to use tools but less likely to use symbols (Lock and Colombo, 1996). Generativity refers to the incredible amount of and variety of human cognitive output – whether it be in the tens of thousands of words in a typical language's lexicon, the almost limitless varieties of song and paintings, or the incredible technological progress that has continued largely unabated from the end of the Middle Stone Age to the present. Finally, the abstract nature of human cognition, similar to what Bickerton (1995) has referred to as "off-line" thinking and what Suddendorf and Corballis (1997) term "mental time travel," strikingly sets humans apart from all other species, which engage largely in the present. While some species can use symbols, only humans can create abstract ones like numbers, words, and religious icons, and it is difficult to conceive of even such advanced creatures as chimpanzees and dolphins as going beyond a simple emotional concept of death or the fulfillment of a current motivationally driven state to such spatially and temporally distant religious concepts as heaven and eternity. Indeed, apes spend the vast majority of their waking lives in immediate, nearby activities (eating and grooming) (see Bortz, 1985; Whiten, 1990), and even Neanderthals

---

[1] In fact, three paintings by a chimpanzee named Congo sold for 12,000 British pounds (over $20,000 US) in 2005 (http://news.bbc.co.uk/2/hi/entertainment/4109664.stm).

appear to have been more constrained in their spatial and temporal mental spheres (Wynn and Coolidge, 2004).

There are two major features that characterize all of the advanced cognitive skills in humans:

1. they all appear to have first emerged between 50,000 and 80,000 years ago, first in Africa and later in Europe and elsewhere; and
2. the context-independent, generative, and abstract expressions of these skills require high levels of a critical neurotransmitter in the brain known as dopamine.

Hence, the emergence of intellectually modern humans around 80,000 years ago arguably represented the beginning of what I will refer to as the "dopaminergic mind." How that mind depends on dopamine, how it came to evolutionary fruition, and the dangers its continued evolution pose for the denizens of industrialized societies in particular will all be discussed in later chapters of this book. First, however, I attempt to refute commonly held explanations (myths) of how human nature evolved. The first myth is that the evolution of human intelligence was primarily a product of genetic selection, while the second is that the specific size, shape, or lateralization of our brain is critical for us to be considered human.

1.1   Myths concerning the origins of human behavior

1.1.1   Was human intelligence genetically selected?

There are many reasons to believe that the origin of advanced human behavior was at least partly controlled by genetic evolution. For one, estimates of the heritability of intelligence, based largely on twin studies that compare the concordance (similarity) of identical twins (which share the same genome) to fraternal twins (which only share the same genetic makeup as regular siblings), are around 0.50 (see Dickens and Flynn, 2001). There are also genetic differences between chimpanzees and modern humans on the order of about 1.2 percent (Carroll, 2003), which in principle could allow for selection for particular genes that may have helped produce the intellectual capabilities of modern humans. Certainly, advanced intelligence should help members of a species survive and reproduce, which according to Darwinian mechanisms should allow that trait to be passed on genetically to offspring. Indeed, it is highly likely that some genetic changes at least indirectly helped to advance human intelligence, although I will argue in Chapter 5 that most of these were probably associated with an overall physiological adaptation that occurred with the dawn of *Homo habilis*.

There are more compelling reasons, though, to believe that *advanced human intellectual abilities are not primarily due to genetic selection*. First of all, genetic expression and transmission have been documented to be modifiable at many levels by a wide variety of influences (especially maternal) that can themselves be passed to offspring in a mode known as "epigenetic inheritance" (Harper, 2005). Indeed, there are ongoing major increases in intelligence (Dickens and Flynn, 2001) and various clinical disorders (Previc, 2007) in the industrialized societies that are occurring despite stable or even opposing genetic influences. For example, the prevalence of autism, characterized by severely deficient social and communication skills, is dramatically increasing despite the fact that most individuals with autism never marry and thereby pass on their genes (see Chapter 4). Second, heritability estimates for intelligence and many other normal and abnormal traits may be overblown because fraternal twins do not share as similar a prenatal environment (a major source of epigenetic inheritance) as most identical twins due to the lack of a shared blood supply (Prescott *et al.*, 1999) and because of the greater similarity of rearing in identical twins (Mandler, 2001). Third, dramatic changes in physiology, anatomy, and behavior are believed to occur when the timing of gene expression is affected by disturbances in key regulatory or hormonal centers such as the thyroid (Crockford, 2002; McNamara, 1995). Fourth, anatomical findings (McDougall *et al.*, 2005) and genetic clock data (Cann *et al.*, 1987; Hammer, 1995; Templeton, 2002; von Haeseler *et al.*, 1996) clearly place the most recent ancestor common to all modern humans at around 200,000 years,[2] yet there is little or no evidence of art, music, religion, beads, bone tools, fishing, mining, or any other advanced human endeavors until more than 100,000 years later (McBrearty and Brooks, 2000; Mellars, 2006; Shea, 2003). One hundred thousand years may not seem like a large amount of time, in that it only constitutes about 5 percent of the total time elapsed from the appearance of *Homo habilis*, but it is more than *ten times longer* than from the dawn of the most ancient civilization to the present. Finally, there is no convincing evidence that genetic factors have played any role whatsoever in one of the most striking of all human features – the functional lateralization of the brain (Previc, 1991).

Although it still remains to be determined exactly how many genes humans actually have, the current best estimate is around 20,000–25,000. Given the 1.2 percent genetic divergence between chimpanzees (our genetically closest living relative) and modern humans, there would first

[2] Genetic clock estimates can be derived from the rates of mutation of various types of DNA (mitochondrial, y-chromosomal, etc.) and the known variations among extant human populations.

appear to be a sufficient amount of discrepant genetic material to account for neurobehavioral differences between us and our nearest primate relation. However, the vast majority of our genome appears to be non-functional "junk" DNA and most of the remaining DNA is involved in gene regulation, with only a tiny percentage of the total DNA (<1.5 percent) actually used in transcribing the amino acid sequences that create proteins (Carroll, 2003). The "coded" sections of the human genome also appear to show less variation between humans and apes than the "non-coded" sections (Carroll, 2003; Mandel, 1996), and much of that difference relates to genes for the protein-intensive olfactory system.[3] In fact, there is no evidence that any proteins, receptors, neurotransmitters, or other components of our basic neural machinery do not exist in chimpanzees (Rakic, 1996). Rather, most of the different genetic sequencing between chimpanzees and humans is in regulatory sections of the genome that affect gene expression (Carroll, 2003), presumably including those that affect brain and body development conjointly. As but one example, there are many genes that affect calcium production, which in turn helps regulate skeletal growth as well as the production of key brain transmitters (see Previc, 1999). Also, there are many genes that can affect the thyroid gland, which has an important influence on body metabolism, body growth, brain activity, and brain size and is arguably a major force for speciation during evolution (Crockford, 2002) and one of the few endocrine structures known to have altered its function during human evolution (Gagneux et al., 2001; Previc, 2002). It is likely, therefore, that changes in regulatory-gene activity and other factors that influence gene expression played some role in the evolution of humans, most probably in its earliest stages (see Chapter 5).[4]

To say that there may have been some influences on gene regulation in humans during the course of our evolution is more defensible than the notion that *specific genes or sets of genes determine advanced human capabilities*. Rarely does a single gene or small set of genes affect a major brain or non-brain disease, and higher cognitive capacities involve even more genes (Carroll, 2003). For example, the combined variance

---

[3] The olfactory system of humans, for example, is believed to express ~500 receptor genes (Ressler et al., 1994), which is much less than other mammalian species that rely on olfaction to a greater extent.

[4] It has recently been claimed that the general mutation rate of genes related to brain growth has increased in humans relative to other primates faster than genes controlling general cellular function (Dorus et al., 2004), but the significance of this preliminary finding is unclear because it is not known whether the genes in question are specific to brain growth as opposed to body growth in general. Indeed, body height correlates with intelligence by roughly the same amount as brain size, and both relationships are subject to environmental influences (Nagoshi and Johnson, 1987; Schoenemann et al., 2000).

accounted for by several key genes known to contribute to intelligence and to various clinical disorders in humans is less than 10 percent (Comings *et al.*, 1996). The polygenic nature of higher cognition is not surprising when one considers the many cognitive skills – discussed in much greater detail by Previc (1999) and in Chapter 3 – that are required for listening to, comprehending, and responding appropriately to a simple sentence such as "Build it and they will come." First, a motor system must recreate in our own minds what is being said; second, an incredibly rapid auditory processor must decode a multitude of acoustic transients and phonemes; third, a capability for abstraction serves to link the spoken words to their correct meaning; fourth, working memory is required to keep the first clause of the sentence in mind as we await the final one; fifth, cognitive flexibility is needed to realize that, after hearing the second part of the sentence, the first part isn't about construction but expresses a more profound thought; sixth, an ability to judge speaker intent aids in further recognizing that this sentence as spoken by a particular individual (e.g. a philosopher) is not about construction; and finally, there must be an ability to assemble and correctly sequence a collection of phonemes that provides a spoken response that we (or any other individual) may have never uttered before. Despite all of this, some researchers such as Pinker and Bloom (1990) have postulated that a single gene or small set of genes may have mutated to create specific language capabilities (e.g. grammar) only found in humans. Indeed, there was great excitement among the scientific world that a "grammar gene" had been identified in a small English family of supposedly grammar-deficient individuals (Gopnik, 1990), who were later shown to have a mutation of a gene known as "Foxp2" (Lai *et al.*, 2001). There eventually turned out to be several major problems with this finding, however. The first was that the affected family members did not have a selective loss of grammar but rather exhibited many other language problems as well as severe speech articulation difficulties, an inability to carry out simple facial gestures (like winking), behavioral disorders such as autism, and even nonlinguistic cognitive deficits (their average *nonverbal* intelligence quotient was found to be only eighty-six, or fourteen points below their unaffected relatives) (Vargha-Khadem *et al.*, 1995). Moreover, the Foxp2 gene mutation turns out not to be associated with the deficits exhibited by most individuals with specific language impairments (Newbury *et al.*, 2002), nor does the human Foxp2 gene resemble that of other species (e.g. avians and dolphins) who possess advanced vocal communication skills (Webb and Zhang, 2005). The final factor mitigating the importance of the Foxp2 gene in human linguistic evolution comes from a recent DNA finding in

Neanderthals, from whom the ancestors of modern humans diverged nearly 400,000 years ago. At least one of the two major variants of the modern human Foxp2 gene relative to that of chimpanzees was once thought to have occurred as recently as 10,000 years ago (Enard *et al.*, 2002), or long after the emergence of the common human genome. However, an analysis of the DNA of Neanderthals shows that they, too, possessed both modern human Foxp2 variants (Krause *et al.*, 2007), which indicates that these variants must be at least 400,000 years old given the estimated date of divergence of the Neanderthal and modern human lineages (Chapter 5).

Another phenomenon tied to the evolution of humans is the lateralization of the human brain for advanced cognitive functions. Two of the most well-known manifestations of cerebral lateralization are the overwhelming and universal preponderance of right-handedness in humans – about 85–90 percent of individuals in Western societies exhibit some form of right motor dominance – and the greater likelihood of suffering serious speech and language deficits (known as aphasias) following damage to the left hemisphere in adulthood.[5] Although brain lateralization of some sort or another is common in the animal world, the degree of functional lateralization of the human brain is remarkable compared to that of other mammalian brains and especially that of the chimpanzee. Indeed, one of the great triumphs of modern neuroscience was the demonstration, mainly through studies of "split-brain" patients in which the connections between the hemispheres were severed to relieve epilepsy (Gazzaniga, 2005), that the left and right hemispheres of most humans not only differ in their linguistic capabilities but also possess very different personalities (the left is more active, controlling, and emotionally detached while the right is more earthy and emotional) and intellects (the left is more analytical, abstract, and future-oriented while the right one is more concrete, better at judging emotional and mental states, and better at visual manipulations, especially 3-D geometrical ones in body space). Indeed, the cognitive and personality differences between the left and right hemispheres of most humans are greater than between almost any two humans, and the specialized functions of the left hemisphere arguably render it almost as dissimilar to those of the right hemisphere as human intellectual functions in general differ from chimpanzees.[6]

---

[5] It is generally accepted that about 90–95 percent of right-handers and about 70 percent of left-handers possess left-hemispheric dominance for speech (see Previc, 1991).
[6] Indeed, Gazzaniga (1983) has even gone so far as to describe the cognitive skills of an isolated right hemisphere as "vastly inferior to the cognitive skills of a chimpanzee" (p. 536).

Although many theorists such as Annett (1985) and Crow (2000) have posited that left-hemispheric language dominance is largely determined by a single gene – and despite evidence that, at least in some species, the overall direction of body asymmetry is subject to genetic influences (Ruiz-Lozano *et al.*, 2000) – the evidence is strongly against a genetic explanation for brain lateralization in humans. First, the likelihood of one member of a twin pair having the same hand dominance as the other is no greater for identical than for fraternal twins (Reiss *et al.*, 1999),[7] and speech dominance for monozygotic twin pairs shows a similarly weak concordance (Jancke and Steinmetz, 1994). Second, neither handedness nor speech lateralization (see Tanaka *et al.*, 1999; Woods, 1986) appears to be related to the genetically influenced asymmetrical position of the major body organs such as the heart, which, in any case, is the same in humans as in chimpanzees. Third, there does not appear to be any evolutionary advantage conferred by the typical pattern of left-hemispheric dominance for handedness, as left-handers and right-handers on average do not differ in academic or athletic achievement or any other personality variables (see Hardyck *et al.*, 1976; Peters *et al.*, 2006; Previc, 1991), although there may be very slight deficits for some individuals with ambiguous dominance (Peters *et al.*, 2006).[8] Fourth, the development of cerebral lateralization is heavily dependent on both cultural and prenatal factors. As an example of cultural factors, aphasia following left-hemispheric damage was very uncommon a few centuries ago in Europe when the vast majority of adults were illiterate and not exposed to the left–right reading and writing of Western languages, and right-handedness remains much less prevalent in existing illiterate populations (see Previc, 1991). As an example of prenatal factors, handedness and other forms of motoric lateralization are greatly reduced in otherwise normal infants born before the beginning of the third trimester and are affected by fetal positioning in the final trimester, which may be crucial as a source of early

---

[7] Although a greater concordance between identical twins usually (but not always) implies at least some genetic influence, the absence of a greater identical-twin concordance almost certainly rules out such an influence. In a meta-analysis by Sicotte *et al.* (1999), which did not include the Reiss *et al.* 1999 study, a significantly greater percentage of dizygotic twins was found to be discordant for handedness, but this difference averaged across twenty-eight studies was less than 2 percent (21.03 percent for monozygotic twins versus 22.97 percent for dizygotic twins) and can be easily accountable by the different child-rearing of the two twin types.

[8] Nonright-handedness does appear to be slightly more associated with both extreme giftedness and mental retardation, for largely nongenetic reasons (see Previc, 1996), but handedness certainly does not predict intelligence in the vast majority of humans (Hardyck *et al.*, 1976).

asymmetrical motion experience in bipedal humans (Previc, 1991). Indeed, the entire edifice of human laterality may be based primarily on primordial prenatal (i.e. nongenetic) factors (Previc, 1991).

Finally, the notion that language and language-linked brain lateralization are determined genetically is contradicted by the nature of human language as a very robust behavior that does not depend on a particular sensory modality (e.g. hearing) or motor system (e.g. speech). For example, individuals who cannot speak or move their hands can communicate with their feet, and those who cannot hear or see can use their hands to receive messages. Humans have invented languages dependent on speech sounds but also on manual signs, tactile signals, fundamental (musical) frequencies, visual icons, clicks, whistles, and probably other signals as well, all demanding many of the same skills described above for speech comprehension and production. Moreover, the mechanisms of language have expropriated the same systems used in more basic motor functions such as chewing, hand movements and eye movements, the latter two of which accompany linguistic thought (Kelso and Tuller, 1984; Kingston, 1990; McGuigan, 1966; Previc *et al.*, 2005). And, the fact that speech is housed mostly in the left hemisphere of humans certainly doesn't imply a causal (or more specifically, a genetic) linkage because the loss in early life of the left hemisphere does not affect subsequent language ability in any measurable way (see next section). Indeed, a pure "language" gene/protein would have to be a strange one in that it would have to:

1. affect language at a superordinate level, independent of any particular sensorimotor modality;
2. affect one hemisphere more than another, even though the lateralization process does not appear to be under genetic control and even though language proceeds just fine in the absence of the originally favored hemisphere; and
3. affect no other sensorimotor or cognitive systems, even though these other systems are closely tied to language processing and output and are, in some case, necessary for language to occur.

Needless to say, no pure language gene has been found or is likely to ever be found.

In summary, a direct, major role of direct genetic selection in language and other higher-order cognitive functions is unlikely. This is consistent with the fact that *all major intellectual advances during human evolution proceeded in sub-Saharan Africa* (McBrearty and Brooks, 2000; Previc, 1999), even though ancestral humans had populated wide swaths of Africa, Europe, and Asia nearly two million years ago. If cognitive ability and not physiological and dietary adaptations – which occurred mostly in sub-Saharan Africa, for reasons to be discussed in Chapter 5 – was the major trait genetically selected for, then why were the other regions of the world in which cognitive ability would have also proven beneficial unable to rival sub-Saharan Africa as the cradle of human evolution?

1.1.2   Did our larger brains make us more intelligent?

The second "myth" concerning human evolution – that we got smarter mainly because our brains got bigger – remains very popular, even among researchers in the field. Yet, there are even more powerful arguments against this view than against the genetic selection theory. After all, elephants by far have bigger brains than anyone else in the animal kingdom, yet most would not be considered intellectual giants; conversely, birds have very small brains (hence, the derogatory term "bird-brain"), but we now realize that some bird species (e.g. parrots) actually possess relatively advanced cognitive capacities, such as language, arithmetic, and reasoning skills (Pepperberg, 1990).

Accordingly, most brain scientists accept that a better measure than brain size for predicting intelligence is brain-to-body weight; using this measure, humans fare very well, along with other creatures that we might consider intelligent (chimpanzees, dolphins, parrots). However, there are problems even with this measure, because the lowly tree shrew – a small, energetic creature that was an early ancestor of primates such as monkeys but is hardly noted for its intellectual prowess – ranks above all others in brain-body ratio (Henneberg, 1998). Moreover, the correlation between brain/body size and intelligence in humans has generally been shown to be very modest, with a typical coefficient that is barely more than the correlation between height and intelligence (~0.3) (see Previc, 1999). Since no researchers have claimed that height is *causally* related to intelligence, there is no reason to assume that the equally modest relationship between brain size and intelligence is also causally related. Moreover, when examining the relationship between brain size and intelligence *within* families to control for dietary and other environmental differences that differ among families, the correlation becomes essentially random (Schoenemann *et al.*, 2000). Indeed, there are even among humans of normal body sizes great variations in brain size, ranging normally from 1,000 cc to over 1,500 cc, and some of the most brilliant minds throughout history have actually had estimated brain sizes toward the low end of that range. The Nobel prize-winning novelist Anatole France had a brain size of only 1,000 g – about the same as the human ancestor *Homo erectus*, who lived over a million years ago – and most individuals with microcephaly (extremely small brains) without other associated disorders such as Down's syndrome, growth retardation etc. tend to be of normal intelligence (see Skoyles, 1999, for a review). For example, one well-studied young mother given the moniker "C2" is estimated to have a brain size around 740 cc (at the low end of the *Homo erectus* range), despite an estimated intelligence quotient (IQ) of 112 (above that of the average human). Finally, the importance of brain size to intelligence is dubious from an evolutionary perspective, in that most of the increase in brain-to-body size in humans over the past million years is explained not by an increase in brain size but rather by a *decrease* in the size of our digestive tract that was arguably made possible by the reduced digestive demands associated with the increased consumption of meat and the cooking of plant foods (Henneberg, 1998). Ironically, the average human brain actually shrank over the past 100,000 years or so from 1,500 cc to 1,350 cc (Carroll, 2003), despite the aforementioned explosion of human intellectual capability (see Chapter 5).

Consequently, researchers have used yet another measure that compares the relative size of different structures such as the cerebral cortex – the outer, mostly gray mantle or "bark" of our brains, on which most of our higher-order cognitive capacities depend (hence, the positive connotations of having lots of "gray matter") – relative to the brain distribution for an insectivore such as the tree shrew. By this measure, also known as the progression index, the human neocortex is by some accounts effectively 2.5 times larger than the chimpanzee's relative to the rest of the brain (Rapoport, 1990), although this has been disputed (Holloway, 2002). Even more strikingly, the area of the neocortex associated with mental reasoning – the prefrontal lobes – occupy 29 percent of the neocortex of humans but only 17 percent of that of the chimpanzee (Rapoport, 1990), although a more recent review argued for no difference in relative frontal-lobe sizes between humans and other apes (Semendeferi *et al.*, 2002). At first glance, this suggests that the size of at least one portion of our brain may indeed account for the intellectual advancement of humans relative to the great apes.

However, the attribution of human intelligence to a larger neocortex or larger prefrontal cortex in particular is as erroneous as the notion that overall brain size is relevant to human intelligence. For one, an individual by the name of Daniel Lyon who had a brain of only 624 cc but essentially normal intellect is believed to have had an especially small cerebral cortex relative to the size of the rest of his brain (Skoyles, 1999). Moreover, research has shown that removal of only the prefrontal cortex in infancy produces no long-term deficits in intelligence in monkeys

(Tucker and Kling, 1969). In fact, removal of major portions of the left hemisphere in infancy produces remarkably few long-term linguistic and other intellectual decrements (de Bode and Curtiss, 2000), even though in normal adult humans the left hemisphere is the predominant hemisphere for most linguistic and analytic intelligence. One child who received a hemispherectomy even ended up with above-average language skills and intelligence, despite the fact that his entire left hemisphere had been removed at the relatively late age of five-and-a-half years and he had disturbed vision and motor skills on one side of his body due to the surgery (Smith and Sugar, 1975). Even more striking is the finding of Lorber (1983), who described many cases of children with hydrocephalus, which occurs when overproduction of cerebrospinal fluid in the center of the brain puts pressure on the cerebral cortex and, if left untreated, eventually compresses and damages it. In one of Lorber's most dramatic cases, a child with only 10 percent of his cerebral cortical mantle remaining eventually ended up with an overall intelligence of 130 (in the genius range), with a special brilliance in mathematics (Lorber, 1983). Somewhat cheekily, Lorber went so far as to entitle his famous chapter "Is your brain really necessary?"

The failure of large-scale cortical removal in infants and young children to substantially affect subsequent intelligence is complemented by the dramatically different intelligences that exist for similarly sized brains. The most striking example of this involves the left and right hemispheres, which are almost identical in weight and shape, aside from a few minor anatomical differences that appear to have little functional significance (see Previc, 1991). Yet, as noted earlier, it would be hard to imagine two more different intellects and personalities. The left hemisphere is impressive at abstract reasoning, mathematics, and most language functions, yet it has difficulty in interpreting simple metaphors and proverbs, in judging the intent of others, and in performing other simple social tasks. By contrast, the right hemisphere is poor at most language functions (it has the grammar of a six-year-old and the vocabulary of an eleven-year-old) and does poorly on logical reasoning tests, yet it is superior to the left hemisphere at proverb interpretation, understanding the intent of others, self-awareness, emotional processing, social interaction, certain musical tasks, and 3-D geometry (Gazzaniga, 2005). Another important example of the stark contrast between an anatomically normal brain and severe abnormalities in higher mental functioning involves the disorder known as phenylketonuria. In this genetic disorder, the enzyme phenylalanine hydroxylase is absent and unable to convert phenylalanine to tyrosine (the precursor to the neurotransmitter dopamine), resulting in a buildup of pyruvic acid and a decrease in tyrosine (as well as dopamine). Because these problems only emerge after birth, when the basic size and shape of the brain has already been established, the brains of persons suffering from PKU appear grossly normal, even though those with PKU suffer severe mental retardation if their excess phenylalanine is not treated by dietary restrictions.

In conclusion, there are compelling reasons to reject as myth the standard view that the evolution of human intelligence and other advanced faculties was determined by direct genetic influences that conspired to change the size and shape of the human brain. On the basis of his own findings with hydrocephalic children, Lorber (1983: 12) concluded that there is an "urgent need to think afresh and differently about the function of the human brain that would necessitate a major change in the neurological sciences." Unfortunately, the revolution in the perspective of the neuroscientific community at large has yet to occur.

1.2 The evolution of human intelligence: an alternative view

1.2.1 Dopamine and advanced intelligence

The pervasiveness of the myth that the ability of humans to think and create in advanced ways is dependent on the overall size of our brains is surprising in that few of us would automatically conclude that a bigger computer is a better one. Indeed, some of the massive early computers had less than one-billionth the speed and capacity of current notebook computers. Rather, it is how the system works – a collective product of such functions as internal speed, amount of parallel processing etc., known as its "functional architecture" – that largely determines its performance. In fact, by any stretch of the imagination, our brain is far larger than it would ever have to be to perform the advanced intellectual functions that we evolved. For example, the number of nerve cells in it (100 billion, as a generally accepted estimate) times the average number of connections per nerve cell (10,000, another generally accepted estimate) times the number of firings per second (up to 1,000, for rapidly firing neurons) allows our brain to perform a comparable number of calculations (10^{18}) as our very best state-of-the-art computers. While using but a fraction of their hardware capabilities, such computers can crunch massive numbers in microseconds, generate real-world scenes in milliseconds, understand the complex syntax of language, and play chess better than the greatest of humans.

What, then, is the essence of why we are so intelligent relative to the rest of the animal world, and especially other primates? Perhaps the most important clue – indeed, the "Rosetta Stone" of the brain – lies in the differences between the left and right hemispheres of the human brain.[9] As already noted, it is our left hemisphere and its grammatical, mathematical, and logical reasoning skills that most obviously differentiates our intellectual capability from that of chimpanzees, despite the comparable size and overall shape of its right hemispheric counterpart and the lack of a known gene underlying its advanced intellect. The right hemisphere is marginally heavier and the left hemisphere may have slightly more gray matter, but the left–right differences are far smaller than between brains of two different humans. There also typically exist larger right frontal (anterior) and left occipital (posterior) protrusions, as if the brain was slightly torqued to the right, as well as some differences in the size, shape, and neural connections of the ventral prefrontal and temporal-parietal prefrontal regions of cortex that house, among other things, the anterior and posterior speech centers (Previc, 1991). However, there is evidently no functional significance to the torque to the right, and the other changes do not appear to be causally linked to the functional lateralization of the brain, since they generally occur *after* the process of cerebral lateralization is well underway (see Previc, 1991).

A much more likely candidate for why the left and right hemispheres differ so much in their functions is the relative predominance of four major neurotransmitters that are used to communicate between neurons at junctures known as the synapses. The four most important lateralized neurotransmitters are dopamine, norepinephrine, serotonin, and acetylcholine.[10] On the basis of a wide variety of evidence (see Flor-Henry, 1986; Previc, 1996; Tucker and Williamson, 1984), it is generally accepted that dopamine and acetylcholine predominate in the left hemisphere and that norepinephrine and serotonin predominate in the right hemisphere. The latter two transmitters are heavily involved in arousal, which explains why the right hemisphere is generally more involved in emotional processing. So, that leaves dopamine and acetylcholine as the two most likely candidates for understanding why the left hemisphere of humans has evolved such an advanced intellect, with at least five suggestions that it is dopamine rather than acetylcholine that

---

[9] The Rosetta Stone was a tablet that contained Ancient Greek and Coptic letters as well as Egyptian hieroglyphics, discovered during Napoleon's occupation of Egypt in 1799 and initially translated by the French linguist Champollion. By matching the different symbol sets, Champollion was able to decipher the previously mysterious hieroglyphic language and in so doing provided an important (if not *the* most important) clue for the understanding of ancient Egyptian culture and society.

[10] The lateralizations of two other important neurotransmitters – glutamate and gamma-aminobutyric acid, or GABA – are not as well-studied and are less interesting in any case from an evolutionary standpoint, for reasons to be discussed later.

underlies human intelligence. Three of these pertain to how dopamine is distributed in the brain, and the other two of these pertain to the known role of dopamine in cognition.

The three major arguments for a role of dopamine in advanced intelligence are:

1. dopamine is highly concentrated in all nonhuman species with advanced intelligence;
2. only dopamine has expanded throughout primate and hominid evolution;[11] and
3. dopamine is especially rich in the prefrontal cortex, the single-most important brain region involved in mathematics, reasoning, and planning.

Nonhuman brains of other advanced species such as parrots, dolphins, and primates differ in many fundamental respects, but they are similar in that all have relatively high amounts of dopamine (Previc, 1999). Most striking is the tiny overall size and almost absent cortical mass of birds, which nonetheless have a well-developed dopamine-rich striatum (Waldmann and Gunturkun, 1993) and nidopallium caudolaterale that support their impressive mathematical and linguistic competencies. Indeed, the latter structure is considered analogous to the mammalian prefrontal cortex because of its high percentage of input from subcortical dopamine regions and its critical role in cognitive shifting and goal-directed behavior (Gunturkun, 2005). As regards the second argument, the dopaminergic innervation of the cerebral cortex is known to have undergone a major expansion in primates; for example, dopamine is limited to specific brain areas in rodents but is densely represented in certain cortical layers in all regions of the monkey brain (see Gaspar et al., 1989). Moreover, the dopaminergic expansion has continued into humans, judging from the almost two-fold increase (adjusted for overall brain size) in the size of the human caudate nucleus – in which dopamine is most densely concentrated – relative to that of the chimpanzee caudate (Rapoport, 1990). By contrast, there is no evidence that any other transmitter has expanded as much, if at all, during human evolution, and the cholinergic content of the human cerebral cortex may have actually diminished (Perry and Perry, 1995). Finally, dopamine appears to be especially well-represented in the prefrontal cortex of humans, and

[11] The term "hominid" has traditionally referred only to those higher primates whose lineage directly led to modern humans, but more recently this term has been extended to apes and a new term – "hominin" – has been devised to apply to the specifically human lineage. The traditional use of "hominid" will be used throughout this book, however.

chemical removal of dopamine from an otherwise intact prefrontal cortex essentially duplicates all of the intellectual deficits produced by outright damage to this region (Brozoski *et al.*, 1979; Robbins, 2000). One major feature of the prefrontal cortex is its ability to recruit other cortical regions in performing parallel mental operations, which is likely to be dopaminergically mediated because dopamine is well-represented in the upper layers of the cerebral cortex, where the connections to the other cortical regions mostly reside (Gaspar *et al.*, 1989; Chapter 2).

Two other reasons linking dopamine with advanced intelligence are its direct roles in normal intelligence and in the intellectual deficits found in clinical disorders in which dopamine levels are reduced. Far more than acetylcholine, dopamine is involved in six major skills underlying advanced cognition: motor planning and execution; working memory (which allows us to engage in parallel processing because we can process and operate on different types of material at the same time); cognitive flexibility (mental shifting); temporal processing/speed; creativity; and spatial and temporal abstraction (see Chapter 3 for a greater elucidation). Working memory and cognitive flexibility are considered the two most important components of what is known as "executive intelligence," and tasks that assess it have been shown by brain imaging to directly activate dopamine systems in the brain (Monchi *et al.*, 2006). Enhanced parallel processing and processing speed, which modern computers rely on to achieve their impressive processing power, are particularly associated with high general intelligence in humans (Bates and Stough, 1998; Fry and Hale, 2000), as are dopamine levels themselves (Cropley *et al.*, 2006; Guo *et al.*, 2006). Second, dopamine is arguably the neurotransmitter most involved in the intellectual losses in a number of disorders including Parkinson's disease and even normal aging, phenylketonuria, and iodine-deficiency syndrome (see Chapter 4). In phenylketonuria, for example, the genetically mediated absence of the phenylalanine hydroxylase enzyme prevents the synthesis of tyrosine, an intermediary substance in the synthesis of dopamine by the brain (Diamond *et al.*, 1997; Welsh *et al.*, 1990). Neurochemical imbalances rather than neuroanatomical abnormalities are believed to be the major cause (and basis of treatment) for almost every brain-related psychological disorder. In particular, either *elevated or diminished dopamine* contributes in varying degrees to Alzheimer's disease and normal aging, attention-deficit disorder, autism, Huntington's disease, iodine-deficiency syndrome (cretinism), mania, obsessive-compulsive disorder, Parkinson's disease, phenylketonuria, schizophrenia, substance abuse, and Tourette's syndrome, all of which are associated with changes in cognition, motor function, and/or motivation (see Previc, 1999; Chapter 4).

The importance of dopamine to these disorders partly explains why dopamine is by far the most widely studied neurotransmitter in the brain. For example, in the exhaustive *Medline* database listing studies involving different neurotransmitters, dopamine was the subject of over 60,000 brain articles through 2008, whereas serotonin – the next most-studied neurotransmitter and a substance that has important inter-actions with dopamine – was the subject of about 38,000 brain articles and acetylcholine (the other predominant left-hemispheric transmitter) was the subject of less than 17,000 brain papers.

1.2.2   The rise of dopamine during human evolution

If the principal reason for the uniqueness of human intellectual and other behavior is that the neurochemical balance in our brains favors dopamine, the remaining great question is how dopamine ended up being so plentiful in the human brain, especially in its left hemisphere. Certainly, there are no new genes that appeared in humans that control the production of the major dopamine synaptic receptors (Previc, 1999), and no variation in these genes within humans seems to strongly cor-relate with variations in intelligence (Ball *et al.*, 1998). As will be dis-cussed further in Chapter 5, it is more likely that dopamine levels may have been indirectly altered by genetic changes that affected cal-cium production, thyroid hormones, or some more general physiological mechanism or adaptation, given that both calcium and thyroid hor-mones are involved in the conversion of tyrosine to dopa (the immediate precursor to dopamine) in the brain. The stimulatory effect of thyroid hormones on both skeletal growth and dopamine metabolism, as well as the stimulatory effect of dopamine on growth hormone, are especially attractive mechanisms for explaining the triple convergence of intelli-gence, brain size, and skeletal height. The importance of the thyroid hormones is also suggested by the previously noted finding that elevated levels of thyroid hormone in humans represent the first con-firmed endocrinological difference between chimpanzees and humans (Gagneaux *et al.*, 2001).

As reviewed by Previc (1999, 2007), there are even more plausible *nongenetic* explanations for why dopamine levels increased during human evolution. As will be discussed further in Chapters 4 and 5, the most likely candidate for a nongenetic or epigenetic inheritance of high dopamine levels is the ability of maternal factors – specifically, the mother's neu-rochemical balance – to influence levels of dopamine in offspring. Not only have dopaminergic systems been shown to be influenced by a host of maternal factors, but there is also compelling evidence that the neurochemical balance of humans has, in fact, changed in favor of increasing dopamine due to a combination of five major factors:

1. a physiological adaptation to a thermally stressful environment (which requires dopamine to activate heat-loss mechanisms);
2. increased meat and shellfish consumption (which led to greater supplies of dopamine precursors and conversion of them into dopamine);
3. demographic pressures that increased competition for resources and rewarded dopaminergically mediated achievement-motivation;
4. a switch to bipedalism, which led to asymmetric vestibular exposure in fetuses during maternal locomotion and resting and ultimately elevated dopamine in the left hemisphere of most humans (see Previc, 1991); and
5. major increases in the adaptive value of dopaminergic traits such as achievement, conquest, aggression, and masculinity beginning with late-Neolithic societies.

The link between bipedalism and brain lateralization exemplifies how epigenetic transmission could have become a seemingly permanent part of our inheritance during our evolution. Bipedalism, together with asymmetric prenatal positioning, creates asymmetrical gravitoinertial forces impacting the fetus, which may in turn create asymmetrical vestibular functioning and neurochemical differences between the hemispheres (Previc, 1991, 1996). Although the ultimate cause of the neurochemical lateralization may be nongenetic, the switch to bipedalism was a permanent behavioral fixture so that the resulting cerebral lateralization would continue for all future generations of humans and superficially appear as it if had become part of the genome itself.

Before addressing the changes in human brain dopamine levels during evolution and history in Chapters 5 and 6, it will first be necessary to further detail the nature of dopaminergic systems in the human brain (Chapter 2) and dopamine's role in normal and abnormal behavior (Chapters 3 and 4). Finally, Chapter 7 will discuss the consequences of the "dopaminergic mind" not only for humans but for other species. As part of that critique, the impressive accomplishments of the dopaminergic mind throughout history will be weighed against the damage it has caused to itself, to others, and to the Earth's ecosytems. It will be concluded that for humans to prosper (and perhaps even survive), the dopaminergic imperative that propelled humans to such great heights and more than anything else defined us as humans must in the end be relinquished.

2   Dopamine in the brain

In this chapter, I will attempt to briefly describe what dopamine consists of, where it is located in the brain, and what basic actions it has. Much of this information will be applied in Chapters 3 and 4 with reference to the roles of dopamine in normal and abnormal behavior.

2.1   The neurochemistry of dopamine

Like all brain neurotransmitters, dopamine is a chemical that contains most of the important building blocks of life – carbon, hydrogen, oxygen, and nitrogen. It is a phylogenetically old transmitter, found in primitive lizards and reptiles existing tens of millions of years ago. The chemical structure of dopamine is shown in Figure 2.1. Dopamine is known as a catecholamine, which derives from its having a catechol group and an amine group that are joined by an additional carbon pair. The catechol group consists of a hexagonal benzene (carbon-bonded) ring with two hydroxyl (oxygen and hydrogen, or OH) groups. The amine group is a molecule comprised of an atom of nitrogen and two atoms of hydrogen (NH_2). The chemical structure of dopamine is not all that special, in that its atoms and molecules derive from some of the most common elements on Earth and especially those found in organic compounds. Carbon, in particular, is essential to life on Earth, partly because it so readily makes bonds with other biological molecules. Another well-known catecholamine that is closely related to dopamine in its chemical structure is norepinephrine (also known as noradrenaline), which is synthesized from dopamine by adding an oxygen atom to one of the hydrogen atoms to form an additional hydroxyl group adjoining the carbon pair that links the catechol and amine groups (Figure 2.1).[1] A third major neurotransmitter

---

[1] In this book, "norepinephrine" will be used instead of the less widely used term "noradrenaline;" however, "noradrenergic" will be used as the adjectival form of "norepinephrine," as is again the custom.

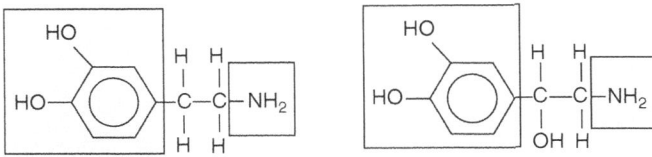

Figure 2.1 The chemical structure of dopamine (left) and norepinephrine (right).

that has an amine group but lacks the catechol structure is serotonin (also known as 5-hydroxytryptamine, or 5-HT), which is considered part of the indoleamine class. Other important neurotransmitters like acetylcholine and glutamate have neither a catechol nor an amine group. A good review of the neurochemistry of dopamine and other neurotransmitters is provided in Cooper *et al.* (2002).

The exact chemical structure of dopamine (or any other neurotransmitter) is not necessarily indicative of what it does in the brain. For example, despite their nearly identical chemical structures and overall stimulant effects on behavior, dopamine and norepinephrine have very different behavioral and physiological effects, and often (and perhaps even mostly) inhibit each other's actions, as will be discussed later. Similarly, apomorphine mimics the action of dopamine in the synapse and also in its behavioral effects (e.g. increasing exploration, motor activity etc.), but it also closely resembles the chemical structure of the drug morphine, which is a powerful analgesic.

Dopamine and all other neurotransmitters engage in a process of neural transmission that in many ways resembles a military campaign (see Figure 2.2). In the region near the dopaminergic cell body, the cell creates the substance dopa from tyrosine (a substance found in proteins) using the enzyme tyrosine hydroxylase, which adds a hydroxyl (OH) group to tyrosine, and the final substance dopamine is then created by breaking apart a carboxyl (C-O-O-H) group and removing the CO_2 from it via the actions of dopa decarboxylase (Step 1). All of this is akin to the assemblage and storage of armaments and other military materiel. Once created, dopamine's movement then resembles a logistical supply line as it traverses along the axon of the neuron and eventually reaches the end of the axon (Step 2). At this point, dopamine is loaded into vesicles until being off-loaded at the membrane of the first neuron (known as the presynaptic membrane) (Step 3). One well-known drug that helps "unload" dopamine at the presynaptic membrane is amphetamine. Once it reaches the presynaptic membrane, dopamine molecules cross the synapse, a tiny fluid-filled space of only

20–40 nanometers (less than 1/10,000th of a millimeter), and quickly contact with the next-in-line neuron. This contact, which is similar to establishing a beachhead, is made at a postsynaptic receptor site, which contains a protein molecule that dopamine attaches to in similar fashion as a key enters a lock (Step 4). Once occupied, a typical receptor then eventually alters the flow of ions into the neurons' membrane and so creates a tiny electrical signal. Dopamine has at least five types of receptors: D_1, D_2, D_3, D_4, and D_5, although the D_1 and D_2 receptors are by far the most numerous and best-studied, with the D_2 receptor appearing to be of greater importance to clinical disorders such as schizophrenia and the D_1 receptor better studied in relation to working memory (Civelli, 2000). Just as in a military campaign, oppositional ("antagonistic") drugs and neurotransmitters can to varying degrees block dopamine transmission by their occupation of dopamine receptor sites on the postsynaptic membrane, thereby preventing stimulation of the postsynaptic cell. Perhaps the best known of dopamine antagonists is the widely used clinical agent haloperidol, while apomorphine is the most prominent of the "agonist" drugs that stimulate the dopamine post-synaptic receptor and are used to mimic the effects of dopamine. Other substances may affect how the dopamine molecule is broken down and rejoined in the presynaptic neuron for yet another attempt to cross the synapse, akin to the military jargon of "falling back and regrouping" (Step 5). A substance that breaks dopamine apart is monoamine oxidase, whose inhibition is used clinically to increase the supply of dopamine and norepinephrine at the synapse. Well-known drugs that block the re-uptake of dopamine back into the presynaptic neuron are cocaine and amphetamine, which render dopamine temporarily in high supply at the synapse, thereby increasing transmission. Once a receptor site on a membrane is occupied, future occupation of that site is often made more effective (or in some cases, ineffective) because of changes in the sensitivity of the membrane (which allows for learning and tolerance to drugs to occur).

Other neurotransmitters may further modulate the activity of dopa-minergic transmission. Generally, well-known transmitters such as serotonin, norepinephrine, and even acetylcholine all alter dopamine release in various regions of the brain, typically by inhibiting it (Gervais and Rouillard, 2000; Previc, 1996, 2006, 2007; Pycock et al., 1975). Indeed, the serotonergic inhibition of dopamine release is arguably the clinically most important neurochemical interaction in the brain (Damsa et al., 2004), particularly in the ventromedial and limbic subcortical and cortical areas (Winstanley et al., 2005), and it will receive special mention in Chapters 3 and 4. Most of the interactions of dopamine with

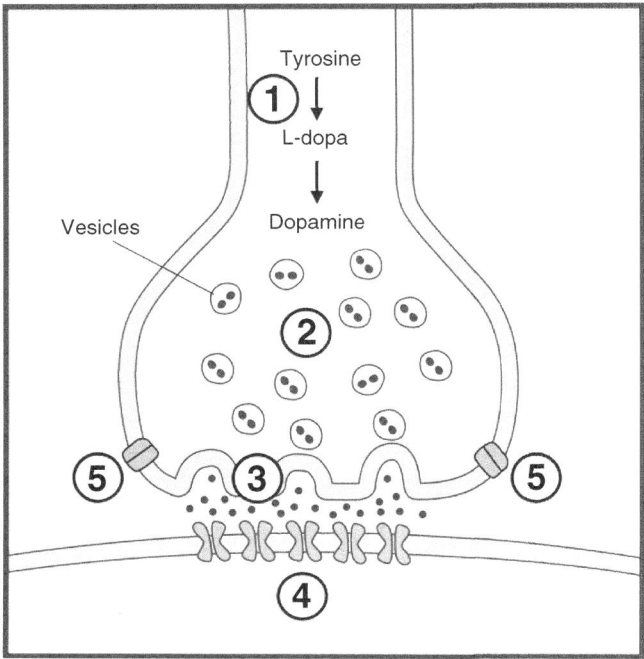

Figure 2.2 The dopamine neuron and synapse: (1) dopamine synthesis; (2) dopamine transport and storage in vesicles; (3) dopamine release from the presynaptic terminals; (4) acquisition of dopamine molecules at dopamine receptors in the postsynaptic neuron; and (5) re-uptake of dopamine back into the presynaptic neuron.

other transmitters probably do not occur at the dopaminergic synapse per se but rather at different synapses on the same neuron. Because of spatial and temporal summation of electrical signals from a variety of synapses on the end-structures (known as dendrites) of the receiving neurons, modulatory influences at nondopaminergic synapses can easily affect the final output of neurons that handle dopaminergic as well as other types of chemical transmission. For example, an inhibitory electrical potential (e.g. from a serotonergic neuron) at one synapse could cancel out an excitatory potential (e.g. from a dopaminergic neuron) at another, thereby preventing excitation of the next neuron in line from occurring.

There is nothing remarkable about the general structure of the dopaminergic synapse shown in Figure 2.2. However, the types of drugs that affect dopamine and their effects on behavior tell us a great deal about dopamine's role in behavior. For example, two substances are especially noteworthy in the synthesis of dopamine – tyrosine (derived from protein-rich food sources such as meat and fish) and phenylalanine, whose failure to be converted to tyrosine partly contributes to the serious mental impairments found in phenylketonuria (see Chapter 4). Tyrosine is then converted to dopa, which is the most effective drug used in the treatment of the severe voluntary motor impairments found in Parkinson's disease, in which the nigrostriatal dopaminergic system in the brain (see next section) suffers massive degeneration. Amphetamine, which enhances the release of dopamine at the presynaptic membrane, generally stimulates behavior and, in low doses at least, improves mental focus, which is why a related drug known as methylphenidate is helpful in treating attention-deficit disorder even as chronic over-use of amphetamine can cause serious psychological problems, including delusions of grandiosity. At the postsynaptic level, antagonists such as haloperidol are known as "typical" antipsychotics in that they have been (and still are) widely used drugs to combat schizophrenia, a hyperdopaminergic mental disorder characterized by delusions and bizarre and disorganized thoughts (i.e. psychosis). Finally, cocaine is a powerful stimulant that prevents the re-uptake of dopamine back into the presynaptic neuron, thereby increasing the supply of dopamine in the synapse. The fact that cocaine and amphetamine are highly addictive substances reveals how some dopaminergic systems are linked to the process of rewards craving (see Chapter 3).

2.2   The neuroanatomy of dopamine

Equally important in understanding what dopamine does in the brain is where it is located. To understand more about its localization, it is necessary to understand a few important points about how the brain is organized and described. The view of the brain in Figure 2.3 shows the three cardinal axes of the brain. The longitudinal axis is known as the anterior(frontal)-posterior one, the vertical axis is known as the dorsal(top)-ventral(bottom) axis, and the lateral axis is directed left-to-right and broken into the outer (lateral) regions or inner (medial) regions. In addition, the brain can be divided into its most superficial portion – termed the cortex, for the Latin word for bark – and the subcortical regions lying below. For example, if one identifies a particular region as being the left dorsolateral frontal cortex, this means it is located towards the upper front of the outer part of the left half (hemisphere) of the brain.

It is also important to recognize that, as shown in Figure 2.3, the cortical and subcortical portions of the brain are each composed of many

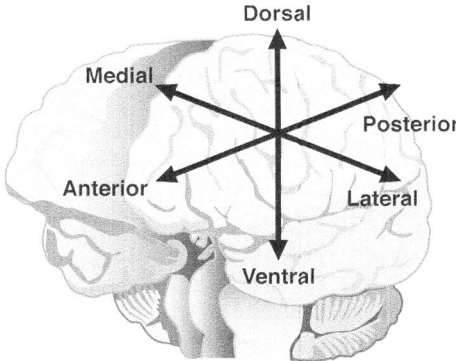

Figure 2.3 The cardinal directions and nomenclature used in brain anatomical localization.

subregions. The neocortex is generally partitioned into four major "lobes:"

1. the occipital lobe, which is located towards the very back of the head and is the only lobe devoted to a single sensory function (vision);
2. the parietal lobe, which is located toward the top back of the head and is involved primarily in spatial relations and the integration of sensory and motor systems in near-body space;
3. the temporal lobe, which is located below a large fissure known as the Sylvian in the lower posterior portion of the brain and is involved in recognition of objects, memory, and language and other functions tied to more distant space; and
4. the frontal lobe, which lies in front of the major fissure known as the Rolandic and is important in motor control and cognition.

A final type of cortex is located medially in the brain and surrounds the thalamus and hypothalamus and corpus callosum, which connects the two hemispheres. This older cortex is known as the limbic cortex and includes such structures as the hippocampus and cingulate gyrus and has major connections to emotion-regulating centers such as the amygdala and hypothalamus. In terms of the subcortex, the most important regions include:

1. the brainstem, containing the midbrain and hindbrain and their centers for basic bodily functions, including sleep and arousal;
2. the diencephalon, which contains the major relay and integration center known as the thalamus (which merges projections of the major sensory and motor pathways with inputs from the arousal centers of

the brainstem) and the hypothalamus (which is located just below the thalamus and maintains most of the hormonal connections with the rest of the body); and

3. the subcortical forebrain, which contains the basal ganglia, long known to be involved in higher-order motor control but also more recently accorded an important role in cognition.

The midbrain and brainstem are of great importance in that they contain the cell bodies that produce a variety of neurotransmitters, including the four major ones: acetylcholine, dopamine, norepinephrine, and serotonin. The projections of these cell bodies can extend all the way to the cortex, although in the case of acetylcholine, the major input to the cortex emanates from the nucleus basalis of Meynert, located in the ventral forebrain. The major noradrenergic and serotonergic systems originate in the locus coeruleus and dorsal raphe nuclei, respectively. The number of neurons directly included in these important neuro-chemical systems is surprisingly small – in humans, only 200,000 carry the serotonergic projections from the midbrain and only about one million carry all the dopaminergic fibers from this region (Rosenzweig et al., 2002). These are, of course, but a tiny fraction of the estimated 100 billion or so neurons in the brain and suggest that major alterations in brain function can, in some cases at least, directly involve only a small percentage of brain cells.

The major dopamine systems originate from several different cell groups near the top of the midbrain in a region known as the tegmentum (covering) (Fallon and Loughlin, 1987; Iversen, 1984). In rodents, two major projections run ventrally from the tegmentum – the nigrostriatal pathways and the mesolimbic. A smaller dopaminergic projection in rodents known as the mesocortical system parallels the mesolimbic pathway but is superseded in primates by reciprocal dopaminergic loops between the dorsal striatum and lateral prefrontal cortex, on the one hand, and between the mesolimbic-linked ventral striatum (con-taining the nucleus accumbens) and the ventromedial prefrontal path-ways, on the other (Cummings, 1995; Mega and Cummings, 1994; Tekin and Cummings, 2002). There are other less-studied dopami-nergic systems that are shorter in length and do not originate in the two classic midbrain sources, such as the tuberoinfundibular system running from the hypothalamus to the pituitary and the projections from the mesolimbic system to the medial preoptic nucleus of the hypothalamus, both involved in regulating physiological interactions with the rest of the body (see Section 2.4). A depiction of some of the major dopaminergic

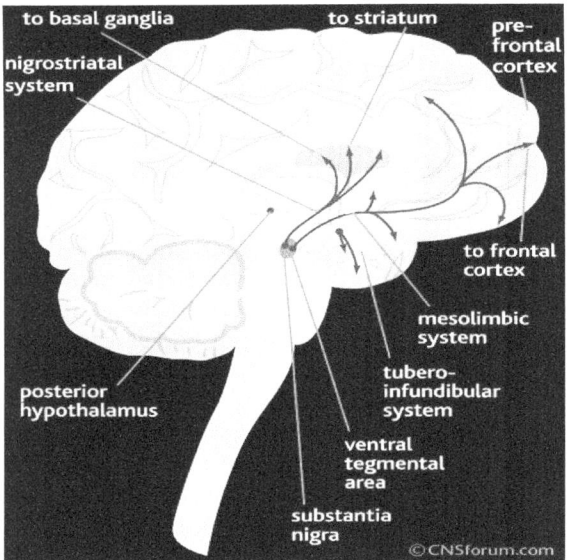

Figure 2.4 Some of the major dopamine systems, shown in a mid-sagittal view. © CNSforum.com.

pathways is contained in Figure 2.4, minus the extensive lateral and medial cortical loops to the dorsal striatal and mesolimbic structures, respectively.

The nigrostriatal pathway will hereafter be mostly referred to as the lateral dopaminergic pathway, because it courses laterally relative to the mesolimbic system. It originates from the A9 subgroup of cells in the tegmentum and links the substantia nigra of the midbrain with the dorsal portion of the corpus striatum, which includes the putamen and caudate nucleus. These structures, along with the globus pallidus, are collectively known as the basal ganglia and have the highest concentrations of dopamine found anywhere in the brain. The nigrostriatal pathways are heavily involved in most types of motor actions and are considered part of the extrapyramidal motor system.[2] As reviewed by Iversen (1984), unilateral damage to the nigrostriatal dopaminergic pathways results in a severe impairment in motor behavior, including reduced motor activity in the limbs opposite (or contralateral) to the side of the brain damage as

---

[2] By contrast, the more direct pyramidal motor system – with cortical axons descending directly into the spinal cord – originates from the giant pyramidal-shaped cells in the primary motor cortex located just anterior to the Rolandic fissure.

well as various postural and turning asymmetries in the direction of the damaged striatum (e.g. rightward turning with right striatal damage). Damage to the lateral dopaminergic system also results in additional higher-order difficulties in the planning, timing, and switching of behaviors (Previc, 1999). Although it is sometimes claimed that the caudate is more entwined with cognitive than motor functions and vice versa for the putamen, brain imaging has shown that both structures are involved in both motor and cognitive functions (Monchi *et al.*, 2006; Mozley *et al.*, 2001). Striatal damage can also create deficits in attending and responding to stimuli (e.g. hemispatial neglect), and the association and prediction of stimuli leading to rewards may also be impaired (Schultz *et al.*, 1997). One indication of the role of the nigrostriatal dopaminergic pathways is what happens when these pathways are severely damaged, as occurs in Parkinson's disease. Persons suffering from Parkinson's disease have a serious inability to perform voluntary motor acts, although they may do remarkably well in more visually elicited, ballistic or arousal-generated motor behaviors. For example, Parkinson's patients may be able to catch a ball if it is suddenly thrown to them (Glickstein and Stein, 1991) or, as in the case of a former Parkinsonian neighbor of mine, move quite well when a swarm of bees start attacking. The nigrostriatal dopaminergic system also has important connections with the prefrontal cortex, particularly its lateral region, which may account for the cognitive deficits found in Parkinson's disease, especially those involving attention, working memory, and shifting strategies (see Previc, 1999). Indeed, disruption of the lateral dopaminergic neurochemical system by substances that destroy dopaminergic neurons in an otherwise anatomically intact prefrontal region mimics the effect of neuroanatomical destruction to this region (Brozoski *et al.*, 1979). One drug that may boost lateral prefrontal activity to a greater extent than activity in other regions is amphetamine (Chudasama *et al.*, 2005), which as already noted helps stimulate the release of dopamine and tends to improve mental focus, at least in low doses.

The mesolimbic pathway, which will hereafter be mostly referred to as the ventromedial or medial dopamine system, emanates slightly more medially from the A10 cell group in the ventral tegmentum and courses medially through the limbic system and limbic cortex. One of the key subcortical structures in this pathway is the nucleus accumbens, which has a high concentration of dopamine and is involved in exploratory, orienting, and motivational/reward behavior. The mesolimbic system also interacts with important medial cortical and subcortical elements, including the amygdala and hippocampus (located adjacent to the medial temporal cortex), the anterior cingulate cortex, olfactory cortex, and the ventromedial frontal cortex. This last region is believed to hold two major subregions that interact extensively with each other – the orbitofrontal frontal cortex, so-named because of its proximity to the eye socket or orbit, and the medial frontal cortex (Ongur and Price, 2000). The extended mesolimbic dopaminergic network is believed to be the one responsible for providing most of our motivational drive and creative impulses and even aggressive behavior (de Almeida *et al.*, 2005), and the most salient behavioral loss following damage to it is a loss of motivational drive. This may be especially true of the ventromedial portion of the mesolimbic system, interacting with the medial shell of the accumbens (Ikemoto, 2007). For example, lack of lever-pressing for food and drink in rats may initially occur following mesolimbic damage (Iversen, 1984), even though the animal is capable of making the proper limb movements (unlike in the case of nigrostriatal damage). The mesolimbic system is also more affected by emotions and stress (Finlay and Zigmond, 1997) and is believed to be the more seriously disturbed in a host of clinical disorders, including schizophrenia, obsessive-compulsive disorder, attention-deficit disorder, and substance abuse. Many of these disorders reflect the inability of the ventromedial and lateral prefrontal cortical centers to inhibit our primitive motivational drives and thought processes. A well-known drug that affects the mesolimbic system preferentially is cocaine (Fibiger *et al.*, 1992), which is highly addictive and, when used chronically, produces compulsive and psychotic behavior resembling schizophrenia (Rosse *et al.*, 1994).

The classic distinction between nigrostriatal dopaminergic involvement in motor control/planning/cognition and mesolimbic dopaminergic involvement in motivational drive and reward is widely held, with considerable justification. In terms of dopaminergically mediated male sexual behavior, for example, damage to the nigrostriatal system impairs the motor aspects of copulation (e.g. mounting of the female) more than precopulatory (motivational) behaviors in the presence of a receptive female, whereas the reverse is true for mesolimbic lesions (Hull *et al.*, 2004). In reality, however, the two systems have extensive cross-talk between them, especially from the mesolimbic motivation regions to the dorsal striatal areas involved in motor programming (Ikemoto, 2007) and they are very heterogeneous, even in nonprimates. Both the lateral and medial frontal regions project to the striatum (albeit to segregated regions), both D_1 and D_2 receptors are prominently located in each system (Civelli, 2000), most drugs interact to varying degrees with both systems, both systems are involved in exploration and behavioral switching, and overactivation of both systems may occur in various psychoses. Moreover, each of them in its own way appears to be involved in distant space, whether it be in providing the basic motivational drive toward distant goals (in the case of the mesolimbic system) or in the prediction of environmental contingencies and the execution of motor strategies required to achieve those goals (as in the case of the nigro-striatal/lateral prefrontal system) (see Chapter 3).

However, the nigrostriatal system is clearly more aligned with the dopamine-rich lateral prefrontal cortex, whereas the mesolimbic system has a greater affinity with medial-frontal cortical structures. This accounts for why combined anatomical damage of the dopamine-rich caudate nucleus and the lateral prefrontal cortex is so devastating for cognition (Tucker and Kling, 1969) and why it is less important than mesolimbic overactivation in obsessive-compulsive and substance-abuse disorders, which involve a derangement of motivational, reward, and inhibitory mechanisms in the brain (Abbruzzese et al., 1995; Adler et al., 2000; Bunney and Bunney, 2000; Rosenberg and Keshavan, 1998; Volkow et al., 2005). The lateral prefrontal cortex may even exert sub-stantial inhibitory control over the mesolimbic system, whether it be in controlling impulsive and compulsive behavior, in controlling the racing thoughts and loosened thought associations in schizophrenia (Bunney and Bunney, 2000), or in dampening the mesolimbic dopaminergic response to stress, particularly in the shell of the nucleus accumbens (Finlay and Zigmond, 1997).

Besides its relevance to explaining various hyperdopaminergic clinical disorders, the relative activation of the medial and lateral dopaminergic systems can explain both the normal variation in dopaminergic per-sonalities, with lateral dopaminergic types being more serious-minded and focused and medial-dopaminergic types being more impulsive and creative (see Chapter 3). It can also explain many altered states such as dreaming, in which the ventromedial dopaminergic system may be unleashed (Solms, 2000), or depersonalization reactions, in which the lateral prefrontal dopaminergic system may be more active (see Sierra and Berrios, 1998), resulting in a state of "thinking without feeling." By contrast, an underactive mesolimbic system can leave an intellectually capable individual lacking in motivation – known as abulia, adynamia, or apathy (Al-Adawi et al., 2000; Solms, 2000; Tekin and Cummings, 2002). This motivational loss is essentially what occurred when prefrontal leucotomies – cuts through the white matter at the base of the frontal lobe – were once performed as accepted practice in the treatment of various neuropsychiatric disorders. Such surgeries have now been replaced by more benign anti-dopaminergic pharmacological treatments that similarly reduce motivation level (Solms, 2000).

Another important facet of the neuroanatomy of dopamine in humans is that dopamine is not confined to a few cortical areas, as it is in lower mammals (Berger *et al.*, 1991; Gaspar *et al.*, 1989). In lower animals, dopamine is located mostly in frontal areas containing primary and association motor cortex and in ventromedial motivational centers such as the anterior cingulate and nucleus accumbens, as befits the important role of dopamine in goal-directed motor activity. In primates and humans, however, dopamine is found in high concentration in most cortical regions as part of the large overall expansion of dopaminergic systems (Gaspar *et al.*, 1989). Still, dopamine is not evenly distributed throughout the brain, even in humans. Dopamine is found in greater concentration in frontal and prefrontal regions and is denser in ventral as opposed to dorsal posterior regions. The anterior-posterior and ventral-dorsal gradients of dopamine have functional significance, in that anterior regions are involved in the voluntary initiation of motor behavior and ventral posterior regions are more involved in attention to distant space, both of which are consistent with the role of dopamine in mediating goal-directed actions toward extrapersonal space (see Chapter 3). By contrast, dorsal parietal-occipital areas of the brain that are involved in visual–manual interactions and consummatory behavior in near-body space have a relatively higher ratio of norepinephrine to dopamine (Brown *et al.*, 1979). Other areas in which dopamine is reduced are posterior brain areas containing the primary sensory representations of the tactile, auditory, and particularly the visual modalities. Indeed, dopamine may actually be inhibited in many parts of the brain during sensory stimulation (Previc, 2006), which is why aberrant dopaminergic phenomena such as hallucinations are more likely to occur during sensory isolation, dreaming, or other external sensory disruptions in which more anterior brain regions begin to "invent" their own percepts (see Section 3.1.3 as well as Previc, 2006). The predominance of dopamine in association cortical areas, in which higher-order sensory processing or cross-modal sensory interactions occur, indicates that dopamine is especially well-suited to making connections among stimuli and events and organizing them into mental plans. This is beneficial in stimulating creativity and in "off-line" thinking and strategizing, important components of abstract reasoning. But, creative thinking divorced from all external reality (e.g. sensory feedback from the body and environment) can also be dangerous, in the form of the bizarre thought associations that excessive dopaminergic activity is known to create (see Chapter 4).

Another important change in the functional role of dopamine from lower mammals to primates is reflected in the different distribution of dopamine across cortical layers. A typical patch of gray-matter cortex in humans has six layers, of which the second from bottom (Layer 5) contains mostly descending projections back to subcortical regions, the third from the bottom (Layer 4) contains mostly ascending subcortical projections to the cortex, and the top layer (Layer 1) contains smaller numbers of cells that have enormous dendritic branching and a large number of connections within and across the entire cortex. The different sizes, neuronal densities, and shapes of these layers in different regions of cortex are the basis for the anatomical classification of different brain areas that, in the widely used scheme of Brodmann, number over fifty (see Rosenzweig *et al.*, 2002). For example, cortical motor regions sending motor commands downstream generally have a large Layer 5, whereas cortical regions involved in sensory processing generally have a large Layer 4 since they receive substantial projections from subcortical relay stations, principally in the thalamus. The major layering difference between primates and most other mammals is the much greater density of dopaminergic neurons in Layer 1 of the former (Gaspar *et al.*, 1989). In rodents, the dopaminergic content of the brain is more uniform across cortical layers, although it is somewhat larger in the lower two levels (which carry a great deal of motor command signals). By contrast, the dopaminergic content of the primate brain, besides being much greater overall, is especially predominant in Layer 1, which has most of the connections with other cortical areas as well as with the striatum and is more involved in coordinating activity during cognitive operations.

It is interesting in this regard that the Layer 1 dopamine neurons have larger branching and connectivity, are more likely to receive inputs from the nigrostriatal pathways, and are more likely not to be co-located with neurotensin (Berger *et al.*, 1991), a neurochemical involved in motivational behaviors such as feeding and drinking. This finding suggests that the less motivationally dependent lateral dopaminergic system may have selectively expanded in humans relative to the mesolimbic one. Another indication of the disproportionate expansion of the lateral dopaminergic system during human evolution is the enormous increase of the human striatum, which contains the highest concentration of dopamine in the brain. As noted in Chapter 1, the striatum nearly doubled from chimpanzees to humans in relative terms (Rapoport, 1990), making it the largest proportionate increase in brain area outside of the neocortex, itself very rich in dopamine.

2.3   Dopamine and the left hemisphere

The distribution of dopamine in humans resembles those of other primates, although it has continued to expand in both relative and absolute terms and is much more highly lateralized than in the case of other primates. Indeed, the lateralization of dopamine activity – greater in the left hemisphere in the majority of humans, particularly in its ventral regions (Previc, 1996, 2007) – may turn out to be the single-most important neurobiological factor accounting for the intellectual abilities and unique personality of that hemisphere. The lateralization of dopamine has been found in postmortem measures of dopamine activity and in lateralized brain activity following ingestion of dopaminergic drugs (Flor-Henry, 1986; Previc, 1996; Tucker and Williamson, 1984, for reviews) as well as more recently in brain imaging studies measuring D_2 binding potential (Larisch et al., 1998; Vernaleken et al., 2007). Greater dopamine in the striatum of one hemisphere induces rotation toward the other side and a contralateral (e.g. opposite) paw preference in animals, which presumably accounts for the predominance of rightward rotation and right-handedness in humans (de la Fuente-Fernandez et al., 2000; Previc, 1991). Dyskinesias (uncontrollable motor outbursts) associated with compensatory excessive dopamine activity after chronic use of dopamine antagonists is greater on the right side (i.e. in the left hemisphere) (Waziri, 1980), whereas motor rigidity in Parkinson's disease (associated with reduced dopamine activity) typically first appears on the left side of the body (i.e. in the right hemisphere) (Mintz and Myslobodsky, 1983). The left hemisphere also predominates in psychological disorders associated with excessive dopamine, such as mania and schizophrenia (see Chapter 4), and it also is more important in dreaming, hallucinations, meditation, and other altered states in which dopamine prevails (see Previc, 2006). Other indirect evidence in favor of a dopaminergic predominance in the left hemisphere is that hemisphere's superiority in voluntary motor skills and its greater grammatical, reasoning, working memory, and other abstract intellectual skills, all of which are dependent on dopamine (Previc, 1999). Also, the left hemisphere in humans may be more important in the initiation of violence (Andrew, 1981; Elliott, 1982; Mychack et al., 2001; Pillmann et al., 1999), although left cortical lesions near the frontal pole may actually increase aggression due to a release in lateral dopaminergic inhibition of the mesolimbic system in that same hemisphere (Paradiso et al., 1996). Conversely, the left hemisphere is deficient in most social and emotional behavior (Borod et al., 2002; Perry et al., 2001; Weintraub and Mesulam, 1983), which can be disrupted by excessive dopamine (see Previc, 2007).

How dopamine evolved to predominate in the left hemisphere of most humans is still not completely understood. However, one leading hypothesis (see Previc, 1996) is that the greater concentration of dopamine in the left hemisphere is an indirect consequence of a greater concentration of serotonin and norepinephrine in the right hemisphere, which as mentioned earlier, serve to inhibit dopamine in the same hemisphere (e.g. Pycock et al., 1975; Previc, 1996). The greater serotonergic and noradrenergic activity in the right hemisphere may result from the predominance of the left vestibular organ and its predominantly contralateral projections that terminate in the right hemisphere (Previc, 1991, 1996). As noted in Chapter 1, it is likely that asymmetry of the otolith organs – which are important in sensing gravity and in maintaining postural control – derives from an asymmetrical gravitoinertial environment in the womb, created by lateralized fetal positioning (Previc, 1991). The otoliths have long been known to influence the sympathetic neurotransmitters norepinephrine and serotonin, because sympathetic arousal is crucial in order to maintain a normal supply of blood to the brain during changes in our body relative to gravity and to aid the body in righting itself during a fall (see Previc, 1993; Yates, 1996; Yates and Bronstein, 2005). The vestibular basis for neurochemical and functional lateralization is consistent with the fact that the absence of normal vestibular function greatly reduces the prevalence of right-handedness, which reflects the greater dopaminergic content of the left hemisphere (Previc, 1996). The vestibular theory of cerebral lateralization is further supported by the importance of vestibular inputs to body-centered perceptual and motor networks that mediate the right hemisphere's greater orientation toward peripersonal as opposed to distant space (Previc, 1998). Whether any other primordial factors influence the lateralization of dopamine to the left hemisphere – along with the major cognitive capabilities it supports – is unclear. As noted in Chapter 1, however, any other causes of brain lateralization are unlikely to be genetic, given the lack of differences between identical and fraternal twins in their concordances for brain lateralization and the lack of any known gene in influencing the direction of cerebral lateralization in humans.

2.4   Dopamine and the autonomic nervous system

One of the most important aspects of dopamine's role in the brain pertains to its involvement with the autonomic nervous system. The two major portions of the autonomic system are the *sympathetic* system, which increases body arousal and metabolism, and the *parasympathetic* system, which generally quiets the body and reduces metabolism (Rosenzweig et al., 2002). Key sympathetic functions, directly regulated by brainstem and hypothalamic structures under the influence of other brain regions and systems such as the vestibular one, include increased cardiac output, vascular constriction (to shunt blood to the muscles and brain), increased oxygen utilization, conversion of fat stores to sugar, and elevated body temperature. By contrast, the parasympathetic system is involved in decreasing cardiac output, decreasing oxygen utilization, increasing vasodilation and heat loss, and reducing digestion.

Dopamine is believed to exert a mostly parasympathetic action in the brain, in that it helps to:

1. lower body temperature, partly by stimulating sweating (Lee *et al.*, 1985);
2. reduce respiration during hypoxia, partly by lowering temperature (Barros *et al.*, 2004);
3. increase peripheral vasodilation, which is important in erectile responses during dopamine-mediated male sexual behavior (Hull *et al.*, 2004); and
4. reduce blood pressure (hypertension) (Murphy, 2000).

In most of these situations, the actions of dopamine are similar to those of acetylcholine but opposite to those of norepinephrine and serotonin, as the latter two neurotransmitters generally serve to increase metabolism, temperature, vasoconstriction, and arousal (Previc, 2004) but tend to inhibit male sexual behavior (Hull *et al.*, 2004; Robbins and Everitt, 1982). The physiological and hormonal actions of dopamine are carried out by various systems, one of which is the tuberoinfundibular pathway into the hypothalamus and the anterior pituitary, which inhibits the release of the female nursing hormone prolactin from the pituitary gland. Other dopaminergic routes involve connections to the medial preoptic nucleus of the hypothalamus (which mediates male sexual behavior as well as parasympathetic thermal and cardiovascular actions) and the vagus nerve (which controls parasympathetic signals to the rest of the body). Dopamine agonists are clinically beneficial in treating a variety of autonomic dysfunctions, including hyperprolactinemia (Gillam *et al.*, 2004), hypertension (Murphy, 2000), and male erectile dysfunction (Giuliano and Allard, 2001). The close physiological linkage between male behaviors and dopamine has its behavioral counterpart in the effect of testosterone to increase dopamine in the brain, in the preponderance of males in many hyperdopaminergic disorders (see Chapter 4), in the incorporation of many masculine behaviors into the dopaminergic personality (see Chapters 3 and 5), and in the masculine behavioral mode (consciousness) of the dopamine-rich left hemisphere (see Chapter 3).

The physiological role of dopamine to dampen physiological arousal and reduce oxygen utilization and metabolism complements its behavioral role in reducing emotional arousal (i.e. increase detachment) and in maintaining motivation and attentional focus in uncertain and stressful environments (see Chapter 3). While dopaminergic systems may mediate some transient positive emotional states such as elation and even euphoria, dopaminergic circuits may be more instrumental in shutting down negative emotional arousal contributing to fear and anxiety by their inhibition of such centers as the amygdala (Delaveau *et al.*, 2005; Mandell, 1980). Behavioral repertoires based primarily on emotional responsiveness would be counterproductive over an extended period of time, just as sustained sympathetic arousal is unsustainable physiologically.[3] Up to a certain point, then, dopaminergic stimulation enables us to maintain control or at least *believe* that we ourselves rather than fate controls our destiny (Declerck *et al.*, 2006), a trait known as "internal locus-of-control" (see Section 3.4.2). However, *too much* dopaminergic stimulation in response to psychological and physiological stress, particularly in the nucleus accumbens and other medial structures, can lead to hallucinations, delusions of grandeur (exaggerated internal control), and even psychosis (Gottesmann, 2002).

It should finally be noted that the role of dopamine (as well as acetylcholine) in facilitating parasympathetic activity is consistent with the well-documented parasympathetic actions of the left hemisphere, in which dopamine and acetylcholine predominate (Previc, 1996). By contrast, the right hemisphere, dominated by noradrenergic and serotonergic activity, is more important in sympathetic arousal and emotional perception and production (see Previc, 1996; Wittling *et al.*, 1998).

2.5   Summary

This brief review of the neurochemistry and neuroanatomy of the major dopaminergic systems clearly demonstrates how important dopamine is to the human brain. In terms of pharmacology, many of the most important drugs to combat psychological disorders have their major effect at the dopaminergic synapse. In terms of anatomy, dopaminergic systems extend throughout the entire cortex and are densest in cortical layers that predominantly connect with other cortical regions. The

---

[3] The metaphor "cooler head" symbolizes the twin roles of dopamine in decreasing physiological arousal (e.g. lower temperature) and promoting analytical thinking; conversely, being "hot-headed" implies more than a mere elevation of cranial temperature.

widespread intracortical projections of the prefrontal dopaminergic system of primates (and especially humans) testifies to its being a major driver of the multitude of parallel processing circuits used in higher mental functions. Dopaminergic dominance is also greatest in regions such as the lateral prefrontal cortex that are known to play a crucial role in higher cognition. The great expansion of dopaminergic systems and their widespread distribution across the cortex in primates reverses the situation in rodents, in which other neurotransmitters are more plentiful and widely distributed across the cerebral cortex and dopamine is confined to mainly motor and prefrontal regions. In addition to the overall expansion of dopaminergic systems during human evolution, the prefrontal/striatal dopaminergic system appears to have expanded relative to the mesolimbic/mesocortical pathways, allowing for the sublimation of more basic and impulsive mesolimbic drives by our rational intellect, although the mesolimbic system continues to play an important role in motivation and creativity in modern humans. The further lateralization of dopamine to the left hemisphere has helped to make that hemisphere the citadel of human reasoning, with abilities far beyond the capability of any other species, as well as the main source of dreaming, hallucinations, and other more chaotic extrapersonal experiences. Finally, dopamine's role in maintaining controlled, goal-directed behavior over emotionally charged behavior, particularly during stress, complements its physiological role in dampening the physiological stress response, which allows highly dopaminergic individuals to function well in extreme environments (Previc, 2004).

3   Dopamine and behavior

The role of dopamine in normal and abnormal behavior has been the subject of a massive amount of research since dopamine was first shown to be linked to Parkinson's disease, schizophrenia, and motivation and reward mechanisms in the 1960s and 1970s. Based on the number of publications that have dealt with it, dopamine is arguably the most important or at least most intriguing neurotransmitter in the brain. It has been implicated in a very large number of behaviors, with a theoretical ubiquitousness that has led it to be facetiously referred to as "everyman's transmitter because it does everything" (Koob, cited in Blum, 1997).

There is one general conclusion regarding dopamine and behavior that a clear majority of neuroscientists would agree on: *dopamine enables and stimulates motor behavior.* Increased dopamine transmission, at least up to a certain point, leads to behavioral activation (e.g. increased locomotion, vocalizations, and movements of the face and upper-extremities) and speeds up motor responses; conversely, diminished dopamine transmission leads to akinetic syndromes, including in the extreme mutism (Beninger, 1983; Salamone *et al.*, 2005; Tucker and Williamson, 1984). One of the distinguishing features of excessive dopamine in the brain is motor behavior known in lower animals as "stereotypy" – constant, repetitive movements that would resemble compulsive behavior in humans (Ridley and Baker, 1982). A dramatic example of dopamine's role in motor behavior is the "stargazer" rat, which has very high dopamine levels and activity levels four to five times those of normals (Brock and Ashby, 1996; Truett *et al.*, 1994). Well-known drugs that increase dopamine transmission such as amphetamine and cocaine are considered stimulants, whereas drugs that block dopaminergic activity such as haloperidol are regarded as major tranquilizers. For example, the dopaminergic agonist quinpirole is known to produce a six-fold increase in locomotion distance in an open-field test (Szechtman *et al.*, 1993), and dopamine is the transmitter most involved in manic episodes in humans, during which activity levels are dramatically elevated (see Chapter 4).

Although degradation of the dopaminergic nigrostriatal system can, if severe enough, affect almost all types of motor activity, the general role of dopamine in stimulating motor behavior belies a considerable degree of specificity in its motor actions. For example, at normal arousal levels, dopamine stimulates:

1. exploratory (seeking) behavior more than proximal social grooming;
2. anticipatory (appetitive) behavior more than consummatory behavior;
3. sexual activity more than feeding;
4. active male sexual behavior (mounting) more than receptive female behavior (lordosis);
5. saccadic (ballistic) eye movements more than smooth-pursuit eye movements; and
6. upward movements more than downward ones.

What do all of the motor behaviors that dopamine selectively stimulates have in common? For a satisfactory explanation, it is crucial to first understand the fundamental role of dopamine in mediating our actions in distant space and time.

3.1   Dopamine and distant space and time

Our interactions in 3-D space are ubiquitous and represent perhaps the single-most important factor in shaping the major pathways of the human brain (Previc, 1998). All of our sensory and motor systems can be linked to various regions of 3-D space, and there are virtually no behaviors (including language) that cannot be linked to particular spatial systems. For example, even language is more closely aligned with auditory and visual systems used in distant space than with tactile and kinesthetic systems involved in reaching activity in nearby space. Likewise, all of our behaviors can be defined by their temporal sphere – i.e. seeking food or a mate or a secondary goal like a good job involves the future whereas touch and the enjoyment of sensual pleasures are associated with the present.

There are four general brain systems that handle our behavioral interactions in 3-D space (see Figure 3.1). One system, known as the *peripersonal*, is mainly used to execute visually guided manipulations in near-body space. This system is biased toward the lower visual field, where the arms usually lie during reaching for and transporting objects to the mouth. Another system, known as the *ambient extrapersonal*, is involved in postural control and locomotion in earth-fixed space and is also biased more toward the lower field, where the ground plane

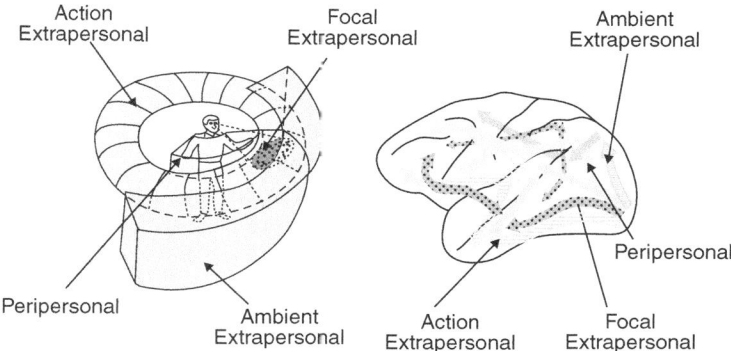

Figure 3.1 The realms of interaction in 3-D space and their cortical representations.

From Previc, F. H. (1998). The neuropsychology of 3-D space. *Psychological Bulletin*, **124**, 123–164, with permission.

predominates. Both of these systems make extensive use of vestibular, tactile and other "body" systems and run dorsally though the parietal lobe and rely on neurotransmitters such as norepinephrine and to a lesser extent serotonin, which are crucial to physiological (and emotional) arousal systems (Previc, 1998). By contrast, dopamine appears to be little involved in these systems or in their representations. As mentioned in Chapter 2, serotonin and norepinephrine are also pre-dominantly localized to the right hemisphere, which is more important both in peripersonal operations and in orienting our postural and per-ceptual systems with gravity.

Dopamine, on the other hand, is the primary neurotransmitter for two systems that deal primarily with extrapersonal space. One of these is the *focal-extrapersonal* system, which is involved in search and scan-ning of the environment and recognition of objects within it. This system operates at a distance because objects are rarely brought into our personal space unless we have already recognized them. This system uses central vision and predominantly saccadic eye movements to carry out the visual search process, and it is localized mainly to the ventrolateral temporal lobe, the lateral prefrontal cortex, and to a lesser extent the region surrounding the parietal eye-fields. It supports detailed visual processing, effortful and controlled interactions with the extrapersonal environment, and the focused executive control func-tions associated with the *lateral dopaminergic* personality/intellect (see Section 3.4.2). A second, more peripherally located system, known as the *action-extrapersonal* one, gives us a sense of "presence" in the environment and aids in exploration, navigation, and orientation to salient stimuli and landmarks. This system extends ventromedially through the medial temporal lobe and hippocampus and limbic areas and on into the ventromedial frontal areas, largely paralleling the medial dopamine cortical system described in Chapter 2. Relative to the focal-extrapersonal system, this system is less focused and concerned with details and it supports the more creative and impulsive functions associated with the *ventromedial dopaminergic* personality/ intellect (see Section 3.4.1).

One consequence of dopamine's association with distant space is that it is important in "exteroceptive" senses such as sight (far vision), hearing, and smell, whereas its counterpart norepinephrine is more associated with "interoceptive" (tactile and bodily) signals (Crow, 1973). Indeed, it is now recognized that there are very close anatomical connections between the olfactory system and the medial dopaminergic motivation systems (Ikemoto, 2007). Another consequence of the dopaminergic association with distant space is that, because the upper field comprises the most distant part of our visual world due to the slope of the ground plane as it recedes from us, the focal- and action-extrapersonal systems are biased toward the upper field, as is dopamine itself. A third correlate of dopamine's involvement with distant space is its involvement with distant (especially future) time. Unlike peripersonal activities, in which the eliciting stimulus or reward may be nearby and requires little effort to reach or procure, pursuit of distant or delayed rewards requires considerable effort and delay, both of which dopamine normally helps to overcome (Denk *et al.*, 2005; Salamone *et al.*, 2005).

The above links to distant space account for why dopamine is selectively involved in a myriad of behaviors that are more likely to be performed outside our immediate space and time. For example, dopamine is critical for:

1. most if not all upwardly directed behavior (Previc, 1998);
2. saccadic eye movements, which are used to explore the distant world and are more upwardly biased than are smooth vergence and pursuit eye movements (Previc, 1998);
3. "male"-type sexual behavior, which in contrast to its female counterpart is more active than receptive and depends more on distant cues such as odors and visual cues than on internal body cues for its elicitation (Robbins and Everitt, 1982).

Conversely, dopamine is less involved in and/or in some cases actually inhibits more peripersonally linked behaviors such as:

1. social grooming and other affiliative behaviors, which are important for group cohesion and for emotional health and are more affected by oxytocin, opioid, and tactile communication (Schlemmer et al., 1980);
2. consummatory aspects of feeding, which are more dependent on opioid (mu-receptor) function (Baldo and Kelley, 2007; Blackburn et al., 1992); and
3. receptive female sexual and maternal behavior such as lordosis (the arching of the back) and lactation that are stimulated by tactile cues and noradrenergic and opioid mu receptors (Depue and Morrone-Strupinsky, 2005; Robbins and Everitt, 1982).

It can even be argued that dopamine's well-documented role in goal-directed behavior (seeking) is a consequence of its primordial orientation toward distant space and time (see next section).

The role of dopamine in 3-D space and time will be further explored in the next few sections. Section 3.1.1 will detail the specific role of dopamine in attending to distant space and time while Section 3.1.2 will describe dopamine's role in goal-directed/reward behavior and Section 3.1.3 will review the role of dopamine in altered states and other experiences in which extrapersonal as opposed to peripersonal themes preside.

3.1.1   Dopamine and attention to spatially and temporally distant cues

A wealth of data has accumulated over the past several decades to show how dopamine in the brain is critical for attending to distant space and time.[1] Dopamine-deficient animals do not easily orient or make associations to distant stimuli (Blackburn et al., 1992), which contrasts with the effects of dopaminergic drugs to increase behaviors controlled by distal cues (Szechtman et al., 1993). Indeed, only animals with normal dopamine levels will attend (orient) to novel distal stimuli if they are engaged in consummatory behavior like feeding (Hall and Schallert, 1988). The effects of dopamine to shift the balance of attention toward distant space and away from peripersonal space is illustrated by its role in hoarding behavior. If an animal is presented with a food object, it can either eat the object on the spot or bring it back to the nest and save it for

---

[1] Distance in this case does not necessarily imply a precise spatial or temporal metric (e.g. meters or minutes) but rather at the very least a departure from the immediate space surrounding the body and the immediate present.

later consumption. Usually hoarding occurs mostly for large, nearby objects (smaller ones can be eaten on the spot), but it can also occur for more distant objects. If an animal is deprived of dopamine, however, the distance for which hoarding occurs is reduced to about two meters (Dringenberg et al., 2000). A similar hoarding failure with distal food objects occurs in animals with damage to the hippocampus (Whishaw, 1993), a key ventromedial structure involved in processing and remembering information from distant space (see Previc, 1998) and part of a more general system that is involved in orientation to distant time (Botzung et al., 2007; Fellows and Farah, 2005; Okuda et al., 2003).

Other examples of dopamine's involvement with distant space are the upward movement and attentional biases produced by dopaminergic activation, which derives from the previously noted association of upward space with the more distant portions of our visual world (Previc, 1998). Examples of these upward biases include rearing on the hindlegs, vertical sniffing, dorsiflexion (raising) of the head, and vertical climbing (as on the walls of the animal's cage). One of the most dramatic examples of such upwardly biased behavior is the nearly continuous dorsiflexion of the head in the hyperactive, dopamine-rich "stargazer" rat (Brock and Ashby, 1996; Truett et al., 1994) (see Figure 3.2a). This behavior has interesting parallels with the upward head and eye movements that have consistently been shown to accompany higher mental activity in humans (Figure 3.2b) and that presumably emanate from the dopamine-rich lateral prefrontal cortex (Previc et al., 2005). The upper-field behavioral biases in animals are further consistent with the upward eye movements and upper-field visual hallucinations that are commonly found in schizophrenia (Bracha et al., 1985; Previc, 2006), which is caused by excessive dopamine. Reduced dopaminergic transmission due to damage to the nigrostriatal pathways or the ingestion of dopaminergic antagonists conversely results in fewer upward eye movements, attentional neglect of the upper field, and nose-diving behavior (somersaulting from high places) (see Previc, 1998, for a review). For example, hypodopaminergic patients with Parkinson's disease tend to reach below the target in memory-dependent reaching (Poizner et al., 1998), and they show fewer upward saccades (Corin et al., 1972; Hotson et al., 1986) and even compress (neglect) the upper field (Lee et al., 2002). By contrast, serotoninergic activation tends to decrease upwardly oriented behaviors (Blanchard et al., 1997), consistent with serotonin's normal inhibition of dopamine in various regions of the brain.

Another aspect of dopamine's involvement in distant space is the well-documented tendency of animals with normal dopamine levels to explore a novel environment by sniffing around its perimeter and poking

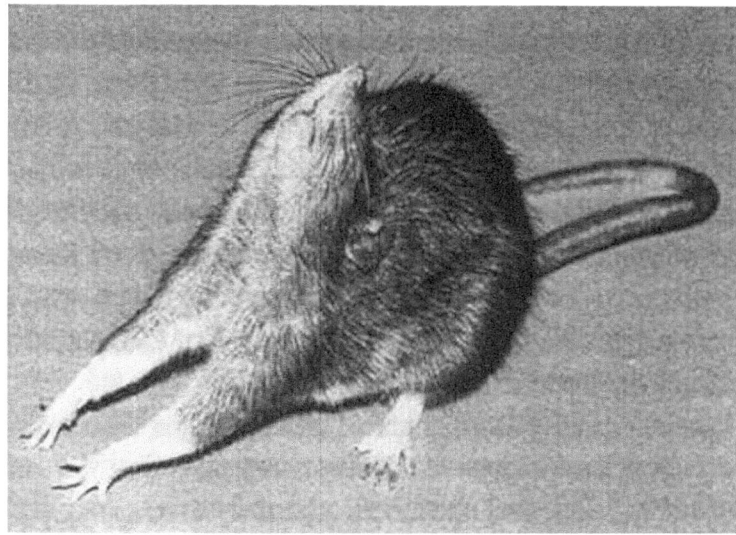

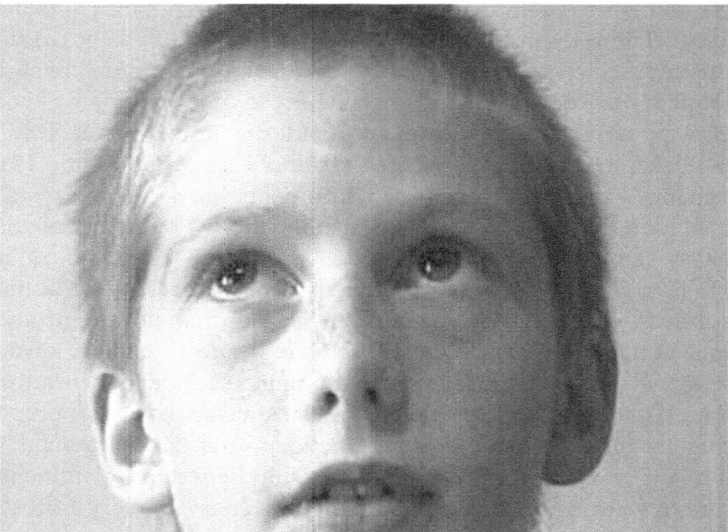

Figure 3.2 Upward dopaminergic biases.
The Stargazer rat (top) from Truett *et al.* (1994). Stargazer (stg), new deafness mutant in the Zucker rat. *Laboratory Animal Science*, 44, 595–599, with permission from AALAS; humans during mental activity (bottom), photo courtesy of Carolyn Declerck.

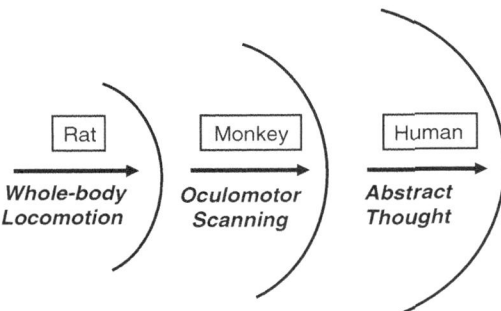

Figure 3.3 The dopaminergic exploration of distant space across mammals.

From Previc *et al.* (2005). Why your "head is in the clouds" during thinking: the relationship between cognition and upper space. *Acta Psychologia*, 118, 7–24, with permission from Elsevier.

their heads in various holes. In fact, normal dopamine levels are essential for animals to prefer a novel to a familiar environment (Fink and Smith, 1980; Ikemoto and Panksepp, 1999; Pierce *et al.*, 1990). By contrast, dopaminergic systems actually diminish exploratory behavior in a familiar environment, as animals deprived of dopamine will continue to explore a familiar environment, and dopamine seems less critically involved in habitual responses (Ikemoto and Panksepp, 1999). The enhanced response to novel environments has led some researchers to conclude that dopamine (in particular, the D_4 receptor) mediates novelty-seeking in humans (Bardo *et al.*, 1996; Cloninger *et al.*, 1993; Dulawa *et al.*, 1999). Unfortunately, novelty-seeking in humans often connotes thrill-seeking or sensation-seeking that elevates arousal, which may actually have little to do with the desire to explore distant environments in a controlled, systematic manner; indeed, sensation-seeking appears to be more dependent on transmitters such as norepinephrine that are more involved in sympathetic arousal (Zuckerman, 1984). It is not surprising, then, that the relationship between dopamine genes and novelty-seeking in humans has not proven to be very robust (Kluger *et al.*, 2002), although at least one measure of exploratory tendency does appear to be correlated with dopaminergically mediated creativity (Reuter *et al.*, 2005).

Even though dopamine may mediate exploration in all mammalian species, the instrument by which the world is explored differs across species (Figure 3.3). In rodents dopaminergically mediated whole-body locomotion and crude distal sensory systems such as olfaction are used to explore a distant world that can be defined by a few meters or tens of meters at most. Primates engage in mostly visual exploration of a greatly expanded distant environment using dopaminergically mediated saccadic eye movements, which are made at a rate of about two to three per second and can locate small food objects at over 25 meters in the distance. In humans, the concept of space goes beyond even the here-and-now, so as to include the ability to imagine even more distant worlds and concepts by means of off-line, abstract thinking and imagination (Bickerton, 1995; Suddendorf and Corballis, 1997). This type of thought is believed to have only recently fully evolved, reaching fruition with the ancient civilizations (see Taylor, 2005; Chapter 6). Despite the vast difference between physically distant space and abstract space, physical and cognitive exploration possess several commonalities, such as dopaminergic involvement and upward biases. Not surprisingly, direct links between spatial foraging and cognitive foraging (Hills, 2006; Hills *et al.*, 2007) and even physical and cognitive number space (Vuilleumier *et al.*, 2004) have been documented in humans.

As for the dopaminergic role in attending to distant (especially future) time, a prime example is its role in reward prediction. Schultz and colleagues (Schultz *et al.*, 1997) have extensively investigated the role of dopamine in reward prediction and have found that dopaminergic neurons in the ventral tegmentum and substantia nigra are active during reward learning. Dopaminergic neurons respond to stimuli paired with rewards during the process of learning, although they stop responding to the reward itself over time and they are actually inhibited if the reward fails to be presented after the cues. Hence, the dopaminergic neurons appear to be sensitive to the predictability and timing of the cue–reward relationship. Moreover, dopamine neurons are not highly responsive even in obtaining rewards if sufficient attention to environmental contingencies is not required (i.e. the reward is too predictable) (Horvitz, 2000; Ikemoto and Panksepp, 1999). That dopaminergic systems are concerned with distant time is also illustrated by the behavior of both rats and humans who prefer a smaller but immediate reward over a larger one that is delayed after receiving drugs that block dopamine synthesis or transmission (Denk *et al.*, 2005; Sevy *et al.*, 2006). The role of dopamine in allowing for delayed reward gratification in animals is consistent with the relatively greater activation of lateral prefrontal regions in humans by decisions involving delayed as opposed to immediate rewards (McClure *et al.*, 2004) and by the effect of lesions of the dopamine-rich ventromedial prefrontal and temporal regions to create what has been termed "myopia for the future" (Bechara *et al.*, 2000; Botzung *et al.*, 2008; Fellows and Farah, 2005; Okuda *et al.*, 2003).

3.1.2   Dopamine and goal-directedness

Goal-directed action has several functional requirements, including:

1. keeping the spatial location and representation of the goal object and landmarks in immediate memory while working toward the goal (working memory);
2. understanding causal relationships in order to predict or control the occurrence of rewards;
3. developing a temporal representation (serial ordering) of the action; and
4. altering behavior on the basis of environmental feedback, as in switching strategies.

Dopamine is the principal neurotransmitter involved in all of these skills (Previc, 1999), and dopaminergic systems – particularly the ventromedial ones – provide the motivational drive for what are often quite complex and effortful actions with delayed gratification. Indeed, as noted in Chapter 2, destruction of the medial dopaminergic systems in the brain can lead to a profound motivational apathy, despite otherwise normal behavior (Salamone et al., 2005; Solms, 2000; Tekin and Cummings, 2002).[2]

The role of dopamine in goal-directed activity would be much more limited if it did not mediate our attention to distant space and time. Predicting reward typically involves the ability to make associations between distant cues and reward objects, and dopaminergic systems have been shown to be especially important when goal-directed activity relies on distal rather than proximal visual or non-visual (e.g. tactile) cues (Blackburn et al., 1992; Szechtman et al., 1993; Whishaw and Dunnett, 1985). And, as already noted, extended behavioral sequences to achieve more distant goals also require an appreciation of the temporal distance to the goal. There even appears to be a close relationship between the extent of one's future consciousness and one's motivation level, with ventromedial frontal lesions diminishing both (see Fellows and Farah, 2005; see also Section 3.4.1).

Dopamine has been widely studied in conjunction with different aspects of goal-directed behavior in animals and in conjunction with addiction, apathy, and other motivational disturbances in humans. The

---

[2] For example, patients receiving medial prefrontal lobectomies and leucotomies in the 1940s to treat psychiatric illness often retained intelligence levels within presurgical limits (Rosvold and Mishkin, 1950), despite profoundly impaired motivational drives.

dopaminergic motivational drive may in many cases be directed toward obtaining rewards necessary for physical survival, such as locating food or water, but in other cases they are not. For example, procurement of sexual rewards, sweet-tasting items, recreational drugs, and rewards with acquired value such as money and knowledge do not satisfy any immediate physiological need. In these cases, the rewards must acquire an incentive value that can justify the sometimes large behavioral expenditures required to achieve them. Most leading theories have posited that dopamine is crucially involved in *incentive motivation* – the motivation necessary to seek and acquire goals/rewards, even if they are not required for immediate physiological survival (Blackburn *et al.*, 1992; Beninger, 1983; Berridge and Robinson, 1998; Depue and Collins, 1999; Horvitz, 2000; Ikemoto and Panksepp, 1999; Salamone *et al.*, 2005).

The specific role of dopamine in performing goal-directed behavior has been studied most in feeding and sexual behavior. Based on the classic distinction between appetitive and consummatory behavior, dopamine is viewed by most researchers as much more involved in appetitive/seeking/foraging behavior (Alcaro *et al.*, 2005; Ikemoto and Panksepp, 1999). Disruption of dopamine usually does not prevent an animal from consuming food that is already nearby, but it does affect its ability to initiate the goal-directed response, learn the behavioral contingencies that lead to the obtaining of the reward, and maintain responding when gratification is delayed (Baldo and Kelley, 2007; Berridge and Robinson, 1998; Blackburn *et al.*, 1992; Dringenberg *et al.*, 2000; Denk *et al.*, 2005; Ikemoto and Panksepp, 1999; Salamone *et al.*, 2005). For example, dopamine neurons are stimulated during the preparatory (goal-directed) behaviors prior to food-ingestion, especially for palatable food, but they are not highly active during the actual consumption of foods, even palatable ones (Baldo and Kelley, 2007; Blackburn *et al.*, 1992). Dopamine is even more critically involved in male sexual behavior, which in the rodent involves a very complex sequence of goal-directed actions, including locomotion to the receptive female, mounting of the female, and copulation. Intact dopaminergic systems, driven by the male hormone testosterone (Hull *et al.*, 2004), are necessary for the first two of these preparatory/seeking behaviors, but they are not necessary for copulation per se (Blackburn *et al.*, 1992). Nevertheless, copulation per se does activate dopaminergic neurons in the nucleus accumbens and other medial structures to a greater extent than does the consumption of regular foods (Blackburn *et al.*, 1992), whose value is more survival-based than incentive-based. As noted earlier, dopamine's facilitation of sexual behavior is primarily confined to males, as most female-type sexual behavior is dependent on other transmitters such as norepinephrine and is actually inhibited by dopamine agonists such as apomorphine (Robbins and Everitt, 1982).[3]

Some researchers have interpreted the role of dopamine in male sexual behavior and other "pleasurable" activities as due to a general involvement of dopamine in hedonic activity. As Berridge and Robinson (1998) note, however, there is a difference between "liking" (actually receiving pleasure from) a reward and "wanting" (craving) it. Liking implies a positive emotional reaction, whereas "wanting" implies a motivational drive that may or may not be associated with a positive emotion or even any emotion at all. A good example of this is addiction to drugs, gambling, sex etc. – the reward itself may be less significant than the process of obtaining the reward. It has repeatedly been shown that medial dopaminergic systems are more involved in "wanting" than "liking" (Berridge and Robinson, 1998; Blackburn et al., 1992; Evans et al., 2006; Ikemoto and Panksepp, 1999; Salamone and Correa, 2002). The greater role of dopamine in motivational drive rather than experiencing sensual pleasure per se may relate to its involvement in obsessive-compulsive and psychologically addictive behavior (see Chapter 4), both of which involve intense behavioral drives for not necessarily pleasurable or beneficial aims. Powerful dopaminergic behavioral drives can be maintained just to avoid aversive events or to obtain incentives (drugs, sex, money etc.) that may over time become very addictive or self-destructive. In the extreme, humans may even choose completely abstract religious and political ideals *over* basic survival. This can be positively expressed in Martin Luther King Jr.'s proclamation that "if a man hasn't discovered something he will die for, he isn't fit to live."[4] But, it is also expressed in the shocking wave of suicide bombings in the Middle East in recent years, the tragic absurdity of the Crusades, and the horrific consequences of a multitude of political and religious ideological conflicts throughout history (see Chapter 6).

A few general statements can be made concerning the role of different dopaminergic systems in motivation and goal-directedness. First, the ventromedial systems in the limbic and basal forebrain areas provide the major motivational drive or impulse for the goal-directed action. The

---

[3] The different characteristics of male and female sexual behaviors ("active" and "receptive," respectively) and the differential involvement of dopamine in them is consistent with other evidence that dopamine is more generally involved in "active" as opposed to "passive" cognitive, attentional and behavioral states (Tucker and Williamson, 1984).

[4] Speech in Detroit, MI, June 27, 1963. Online, available at: www.quotationspage.com/quote/24968.html.

orbitofrontal and anterior cingulate cortices then help to organize and focus the goal-directed behavior into meaningful goal-directed sequences. Ultimately, the lateral prefrontal cortex, in conjunction with striatal systems, regulates these goal-directed actions in terms of prediction and feedback from the environment (e.g. by updating the original target behavior in memory on the basis of current contextual information). The lateral prefrontal network also provides the capacity to shift actions or strategies, which it does by inhibiting previous responses. The involvement of the lateral dopaminergic systems in working memory and cognitive shifting and other "executive"-type behavior ultimately underlies the larger dopaminergic role in higher intelligence, as discussed in Section 3.2.

3.1.3 Dopamine and extrapersonal experiences

One interesting manifestation of the role of dopamine in mediating both upward and distant space and time is its involvement in altered states such as dreaming and hallucinations and in religious experiences in humans. As I recently described (Previc, 2006), all of these activities involve:

1. a preponderance of extrapersonal (distant) sensory inputs (e.g. auditory and visual);
2. a dearth of peripersonal ones (e.g. tactile and kinesthetic);
3. an emphasis on extrapersonal themes (e.g. flying or moving outside of one's body); and
4. the presence of upward eye movements and other upper-field biases.

Dopaminergic activation, caused either by altered physiological states (e.g. hypoxia/near-death, sensory deprivation/isolation, extreme stress) or other neurochemical factors (reduced glutamatergic, cholinergic, and serotonergic activation to varying degrees), appears to be the common denominator in all of these experiences (Previc, 2006). One of the most dramatic manifestations of dopaminergic activation is the out-of-body experience, which can occur during hallucinations, dreaming, sensory deprivation, hypoxia, psychotic disorders, hypnosis, transcendence, and psychologically traumatic events such as rape. Reduced sensory input releases the normal inhibition of serotonergic, cholinergic, and noradrenergic sensory outputs on dopamine, which is poorly represented in sensory processing areas. By contrast, hypoxic "near-death" experiences, dreaming, and depersonalization reactions during traumatic experiences may serve to elevate dopamine as part of a general quieting of the body that conserves oxygen during hypoxic episodes, combats the rise in temperature before sleep (our temperature is lowered during sleep and dreaming), and lowers emotional arousal by reducing the sympathetic hormones involved in the response to psychologically stressful events (see Chapter 2).

Many researchers have posited a fundamental similarity between dreams and hallucinations (Hobson, 1996; Rotenberg, 1994; Solms, 2000), which even share an etymologic root.[5] Hallucinations bear a close relationship with dreaming, as indicated by the fact that the former frequently occur in normal individuals following sleep deprivation or during or just prior to sleep onset ("hypnogogic hallucinations") or just after awakening ("hypnopompic hallucinations") (Cheyne and Girard, 2004; Girard and Cheyne, 2004; Girard et al., 2007). Moreover, hallucinations can even occur in individuals while awake when accompanied by bursts of rapid-eye-movement activity, which is normally associated with dreaming (Arnulf et al., 2000). When ingested before sleep, hallucinogens such as LSD are known to potentiate dreaming (Muzio et al., 1966), and schizophrenics describe their hallucinations as dream-like (Gottesmann, 2002). Dreaming and hallucinations are also closely tied to religious traditions or linked to religious experience (Batson and Ventis, 1982: Chapter 4; Gunter, 1983; Koyama, 1995; Pahnke, 1969; Saver and Rabin, 1997). Dreams have been considered a means of receiving messages from the supernatural (e.g. Joseph's in the Book of Genesis) and meeting ancestors, as exemplified by the construction of many ancient Japanese religious temples in order to foster dreaming (Koyama, 1995). Ingestion of hallucinogenic drugs leads to mystical experiences and religious imagery (see review by Batson and Ventis, 1982: Chapter 4; Goodman, 2002; Pahnke, 1969; Saver and Rabin, 1997), and hallucinations during epilepsy and paranoid schizophrenia are hypothesized to have led to experiences that inspired many of the world's leading religions, including those of St. Paul (the founder of Catholicism), Mohammed (the founder of Islam), and Joseph Smith (the founder of the Mormon religion) (see Saver and Rabin, 1997). Indeed, epilepsy and schizophrenia, both of which involve intense activation of the ventromedial dopaminergic pathways, are two disorders clearly linked to hyper-religiosity (see Previc, 2006), as epilepsy was termed the "sacred" disease by the Greeks (Saver and Rabin, 1997) and "insanity" was originally a Hebrew word referring to those carried away by religious visions (see Previc, 2006).

---

[5] Both derive from "to wander" in some languages – e.g. hallucination from the Latin "alucinari" and dream ("reve" and "reverie" in French) from the French "resver."

In both dreams and hallucinations, serotonergic neurons in the raphe nuclei shut down, helping to unleash dopaminergic activity in normally inhibited medial frontal pathways (Gottesmann, 2002). One interesting difference between dreams and hallucinations is that whereas dreams are most likely to occur in rapid-eye-movement sleep, which is accompanied by cholinergic activation (Hobson *et al.*, 2000), hallucinations are more likely to involve reduced cholinergic inhibition of dopaminergic activity, since anticholinergic drugs like atropine are potent hallucinogens. As noted earlier, one reason why dopamine is so involved in dreams and hallucinations is that dopamine is ordinarily not involved in the basic processing of sensory information but rather is much more plentiful in higher-order perceptual and associative regions. When the primary sensory systems – particularly those providing feedback concerning tactile and kinesthetic sensations – are shut down during sensory deprivation, dreaming, and other altered states, dopaminergic systems begin to make spurious and chaotic associations among internally generated stimuli that can be confused as associations among actual stimuli emanating from the environment (Previc, 2006).

It is interesting to note that during the dopaminergic activation in dreams, anesthesia, hypoxia, hypnosis, meditation, and mystical and delusional states in which out-of-body experiences are common, the eyes tend to roll upward (see Previc, 2006). Moreover, out-of-body illusions and hallucinations like flying are more likely to occur in the upper field and beyond arm's reach (Cheyne and Girard, 2004; Girard *et al.*, 2007), and religious beliefs and experiences in humans are generally biased toward upward, distant space (e.g. heaven) (Previc, 2006). Religious practices such as meditation require a focus on upper space (e.g. focusing on the "third eye," a region on the forehead between the two eyes) or entail a diminution of bodily signals, allowing for transcendence (loss of self). Furthermore, many religious symbols and temples are located in or protrude into upper, even astrological space (e.g. spires and domes, pyramids, mountainside monasteries, sacred mountains, angelic figures, Stonehenge). Similarly, great religious moments and themes are entwined with upper space (Moses on Mt. Sinai, Jesus on the mountaintop, Mohammed transported by the angel Gabriel on a chariot into the sky, Shamans sending spirits into the sky etc.). Positive (approach) elements of religion such as heaven and angels are more likely to be associated with upper space, whereas negative (avoidance) elements such as hell and serpents are more likely to be associated with lower space (Meier *et al.*, 2007; Previc, 2006), in line with a general bias at least in Western cultures to value upper space more highly (e.g. *uplifted* versus *downtrodden*; *exalt* versus *debase*, etc.) (Meier and Robinson, 2004; Previc, 2006).

Finally, religious (or at least spiritual) experiences are usually associated with activation of the ventral brain regions – especially the medial and superior temporal lobe and the medial prefrontal cortex – whereas peripersonal regions such as the posterior parietal lobe appear to be quieted during religious activity (Previc, 2006). The occurrence of symptoms similar to those found in medial-temporal seizures – such as olfactory and visual illusions and feelings of being spatially lost – has been frequently noted in normal persons with paranormal or spiritual beliefs and experiences (e.g. Britton and Bootzin, 2004; Fenwick et al., 1985; MacDonald and Holland, 2002; Morneau et al., 1996; Persinger, 1984; Persinger and Fisher, 1990; Persinger and Markarec, 1987). As already noted, the ventromedial cortical regions are among the most dopamine-rich regions of the cortex, as is consistent with the fact that all major drugs that create mystical experiences – glutamate antagonists, serotonin antagonists, acetylcholine antagonists, and dopamine agonists – ultimately serve to tilt the neurochemical balance toward dopamine (Previc, 2006). Due to its facilitation of associations and predictive relationships among distal stimuli, dopamine is believed to mediate superstitious associations, i.e. beliefs that events are not merely coincidental but are rather causally related. Superstitious behavior is linked to a distorted view of chance, which is a hallmark of paranormal thought (Brugger et al., 1991), and the underestimation of chance and randomness can lead highly dopaminergic persons to overestimate their control over events either directly through one's own ability or indirectly by tapping into spiritual forces through prayer and mediums. The heightened attempt to control events, to be discussed further in Sections 3.4.2 and 3.4.3 in connection with the "locus-of-control" trait and the left-hemisphere style, underlies many religious rituals as well as nonreligious behaviors such as gambling. In fact, both religious rituals and gambling are associated with obsessive-compulsive (heightened control) tendencies and obsessive personality styles (Previc, 2006), which are dependent on activation of the medial dopaminergic pathways (see Chapter 4).[6]

Delusions, hallucinations, and dreaming are relatively more likely to involve activation of the left hemisphere of humans (Previc, 2006). The left hemisphere also appears predominant in most religious experiences

---

[6] Coincidentally (or perhaps not), for many years the only endowed chair in paranormal studies – the Bigelow Chair for Consciousness Studies – was located in the heart of gambling country at the University of Nevada at Las Vegas. It is also worth noting that male sexual titillation, fast-paced excitement, and copious amounts of alcohol that characterize Las Vegas further stimulate the dopaminergic drive and heighten the proclivity to gamble.

and behaviors, albeit to a lesser extent (Previc, 2006). The left-hemispheric predominance in dreams and hallucinations and other altered states runs counter to the common notion that the more intuitive right hemisphere is the source of such states. In reality, the "earthier" right hemisphere is the anchor for our bodily senses, emotions, and sense of self. Damage to the right hemisphere is more likely to produce a neglect of the body (as in dressing apraxia), peripersonal manipulative disorders (as in constructional apraxia), an altered body image and even denial of bodily handicaps (termed "anosognosia"), impaired body orientation in space, reduced self-awareness, and impaired emotional recognition and expression (see Cutting, 1990; Hecaen and Albert, 1978). In extreme cases, damage to the body/self systems of the right hemisphere can lead to actual depersonalization and somatoparaphrenic delusions, in which the patient may attribute sensations on their own body to an alien entity.[7]

Even more specifically, the altered states in which extrapersonal inputs prevail result from activation of the posterior (temporal) and anterior (prefrontal) portions of the ventromedial *action-extrapersonal* system, as opposed to the laterally based focal-extrapersonal system. In essence, ventromedial dopaminergic activation results in the "triumph" of extrapersonal brain activity over the body systems that anchor our self-concept and our body orientation as well as a triumph over the more "rational" executive intelligence maintained in the lateral dopaminergic systems (Previc, 2006). Conversely, the lateral (focal-extrapersonal) dopaminergic systems may be more responsible for the enormous power of the human intellect – particularly its abstract intelligence – as discussed in the next section.

3.2 Dopamine and intelligence

From what has been written above about dopamine's role in goal-directed and exploratory behavior, it would be surprising if dopamine did not play an important role in intelligence. However, the role of dopamine in intelligence clearly goes beyond its motivational role and can be tied to at least six specific cognitive skills, *all of which can be linked to extrapersonal space*. Before describing these skills, I will briefly review the general evidence for a relationship between dopamine and intelligence.

---

[7] A vivid example of the loss of connectedness with one's own self or body following right-hemispheric damage is the tendency to see the other person rather than oneself in morphed images of oneself and a famous celebrity (Keenan *et al.*, 2000). By contrast, the isolated right hemisphere sees the morphed image as more resembling oneself than the well-known person.

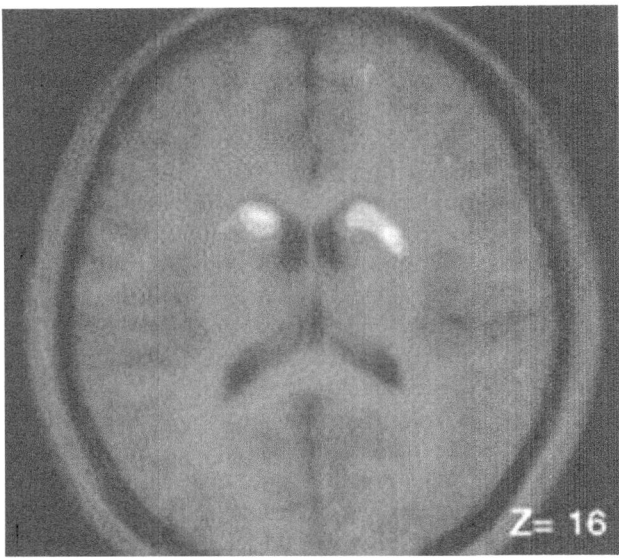

Figure 3.4 An axial (horizontal) section of a human brain showing reduced dopamine D_2 receptor binding (increased dopamine activity) in the left and right caudate nuclei in a reversal shift memory task.

From Monchi *et al.* (2006). Striatal dopamine release during performance of executive functions: A [^{11}C] raclopride PET study. *Neuroimage*, 33, 907–912, with permission from Elsevier.

There are several direct pieces of evidence for a paramount role of dopaminergic systems in intelligence. First, reduced prefrontal dopamine levels are believed to be a prime cause of the working memory and other executive disturbances that occur in phenylketonuria (Diamond *et al.*, 1997; Welsh *et al.*, 1990). As previously noted, loss of just the dopaminergic content of prefrontal cortex reproduces most of the symptoms of outright removal of this region (Brozoski *et al.*, 1979). Second, dopamine levels are diminished due to extreme iodine-deficiency during prenatal brain development because inadequate levels of thyroid hormones limit the conversion of tyrosine to dopa, contributing to widespread mental retardation (Previc, 2002). Third, dopamine binding and other measures have been shown to predict verbal, analytical, and executive intelligence in humans (Cropley *et al.*, 2006; Guo *et al.*, 2006; Reeves *et al.*, 2005). Indeed, dopamine release in the dorsal striatum increases during executive tasks involving cognitive shifting (Monchi *et al.*, 2006), as illustrated in Figure 3.4. Fourth, dopaminergic systems are among the most profoundly affected during the cognitive decline of aging, with the number of dopamine receptors correlating significantly with performance on abstract reasoning, mental flexibility, and a variety of other cognitive tests (Bäckman *et al.*, 2006; Braver and Barch, 2002; Volkow *et al.*, 1998). Finally, neonatal damage to both the prefrontal cortex and caudate nucleus destroys most of the striatal/lateral-prefrontal dopaminergic system and produces permanent cognitive impairments in monkeys (Tucker and Kling, 1969), although this is not as true when one of the structures is spared.

There is also much indirect evidence that supports the crucial role of dopamine in intelligence. From a clinical perspective (see Chapter 4), cognitive function is altered to varying degrees in most diseases in which dopaminergic activity is abnormal, including attention-deficit disorder (where lateral prefrontal dopamine may be insufficient relative to ventromedial dopamine), autism (where lateral dopamine may be excessive), bipolar disorder (where dopamine is elevated during the manic state), Huntington's disease (in which dopamine is relatively elevated in the striatum), Parkinson's disease (where striatal dopamine is reduced due to damage to the nigrostriatal pathways), and schizophrenia (where dopamine in the ventromedial/mesolimbic systems is elevated). The symptom profiles in these disorders are far from identical, which is not surprising given that they affect different dopaminergic systems and to varying degrees other key neurotransmitters. However, one deficit common to virtually all of these disorders is impaired "executive" intelligence (e.g. working memory, planning, shifting strategies), in line with the importance of dopaminergic systems housed in the lateral prefrontal cortex for working memory, cognitive shifting, and other components of fluid and executive intelligence (Braver and Barch, 2002; Diamond *et al.*, 1997; Kane and Engle, 2002; Nieoullon, 2002; Previc, 1999). Other indirect evidence of dopamine's role in intelligence is the relatively high amount of dopamine in nonhuman orders (parrots, cetaceans, and primates) that are believed to possess advanced intelligence, despite their very different brain shapes and sizes (see Previc, 1999).[8] Finally, as noted earlier, the high level of dopamine in the left hemisphere is commensurate with that hemisphere's leading role in abstract reasoning and executive intelligence (see following sections).

Some theorists (e.g. Neioullin, 2002) argue that dopaminergic systems contribute less to standardized intelligence measures than to fluid/

---

[8] Interestingly, at least one of these species (parrots) may also exhibit excessive tics and other stereotyped movements that characterize hyperdopaminergic disorders such as autism, obsessive-compulsive disorder, and Tourette's syndrome in humans (Garner *et al.*, 2003).

executive intelligence. But, even the prefrontal cortex as a whole is not believed to be crucial for scores on standardized intelligence tests (Kane and Engle, 2002; Rosvold and Mishkin, 1950), which typically measure "crystallized" or knowledge-based intelligence to a greater extent than fluid intelligence. Nonverbal intelligence tests such as the Raven's Progressive Matrices Test that measure abstract reasoning are, by contrast, primarily designed to measure fluid intelligence, which underlies operations such as planning, working memory, shifting strategies, and abstract concepts that are more closely related to the concept of "g", or the general intellectual factor (Carpenter *et al.*, 1990; Kyllonen and Christal, 1990). One particular dopaminergically mediated executive skill – working memory, the ability to store items in memory and perform multiple parallel operations on them – is even considered by some researchers to be the single-most important element of "g" (Kyllonen and Christal, 1990).

Of course, dopamine is not the only neurotransmitter involved in intelligence. Acetylcholine is also involved in working memory, attention, and motor programming and is similarly lateralized to the left hemisphere (see also Ellis and Nathan, 2001; Previc, 1999, 2004). Norepinephrine is also involved in working memory and certain motor operations and, since it is derived from dopamine, it is also affected by the failure to develop normal tyrosine and dopamine levels in iodine-deficiency disorder and phenylketonuria. Certainly, the increase in cardiac output and temperature with mental effort (Al-Absi *et al.*, 1997) points to a contribution of the noradrenergically stimulated sympathetic nervous system in mental activity (see Previc, 2004). In contrast to the above neurotransmitters, only serotonin does not appear to play a major role in cognition – indeed, it tends to decrease orienting, vigilance, and working memory (Dringenberg *et al.*, 2003; Luciana *et al.*, 1998; Schmitt *et al.*, 2002), in line with its general oppositional role to dopamine.

As will be described in the following sections, dopamine's contribution to intelligence can be traced to its involvement in six primary cognitive skills: motor programming, working memory, cognitive flexibility, abstract representation, temporal analysis/sequencing, and generativity/creativity (see Previc, 1999). These skills are important not only to general intelligence but also to language, which many theorists consider to be the most advanced talent in humans. Linguistic competence would certainly collapse without these skills, all of which depend in most adults on a normally functioning left hemisphere, and it is also difficult to imagine language *not* being invented by individuals who possess these skills. By contrast, the notion of special linguistic abilities, processing centers, or even genes (e.g. Pinker and Bloom, 1990) that act independently of overall intelligence has yet to be demonstrated. This is not to say that someone with above-average verbal skills may in some cases be deficient in visuospatial abilities and vice versa – obviously, each of us has abilities and/or interests that allow us to perform better in some skill areas than others. But, overall there is a high correlation between linguistic abilities and intelligence (Daneman and Merikle, 1996).

What is important from the standpoint of the larger role of dopamine in brain function is that *all of these skills can be tied directly or indirectly to processing in distant space and time.* Motor programming and planning are necessary for carrying out the sequential actions needed to achieve distant goals, while working memory allows us to receive and hold new information while updating our goal-directed behavior, and cognitive flexibility allows us to change course as we move along toward the distant goal. Abstract representation allows us our cognitive apparatus to escape immediate perceptual images, and rapid sequential processing is needed in processing the complex and rapidly changing signals (particularly auditory ones) in extrapersonal space as we move through it. Finally, the process of generativity/creativity is important in stimulating new associations among various stimuli and responses. That these cognitive skills can be invoked during pure mental thought or when complex goal-directed behavior is required in nearby space does not negate the fact that under natural conditions, humans and other animals would have little need of them if goals were easily obtainable in immediate, nearby space.

3.2.1 Motor programming and sequencing

As already noted, dopamine is crucially involved in voluntary, sequential motor behavior. Dopamine is richly represented in frontal motor areas and is especially involved in motor learning, timing, and programming. Dopaminergic systems in the caudate nucleus are also important in controlling the rhythm of motor outputs (Aldridge *et al.*, 1993), with reduced dopaminergic levels affecting voluntary movements more than reflexive ones and complicated movements more than basic ones. Orofacial movements are particularly facilitated by dopamine, which in excess leads to tics and other orofacial stereotypical behaviors (see Chapter 4). The sequencing of speech is particularly affected by dopaminergic alterations, whether that be overactivation, as in stuttering (Brady 1991), or underactivation, as in Parkinson's disease (Lieberman *et al.*, 1992; Pickett *et al.*, 1998).[9] As previously noted, severe depletion

---

[9] Indeed, spoken language is dominated by the same basic motor rhythm (5 Hz) as are chewing and other orofacial behaviors (Kelso and Tuller, 1984).

of brain dopamine results in the loss of all voluntary motor behavior, including speech.

The notion that motor programming and motor behavior are closely linked to intelligence has been somewhat contentious. Since the early part of the twentieth century, leading behavioral theorists such as Watson and Skinner have viewed linguistic thought as involving motor circuits, especially those involved in speech. Certainly, speech circuits are activated during verbal thought (McGuigan, 1966), and speech bears considerable relation to other orofacial movements in terms of its timing, duration, and musculature (Kelso and Tuller, 1984). Eye movements, usually upward, are also activated during mental arithmetic and other complex tasks, and even relatively simple mental calculations become quite difficult when subjects are forced to look downward or not make any eye movements (Previc et al., 2005). As but another example, topological operations like mental rotation involve activation of the hand-movement representations in the dorsal parietal and premotor regions of cerebral cortex (Cohen et al., 1996; Vingerhoets et al., 2002), which makes sense in that unusual rotations of objects primarily occur in peripersonal space, in conjunction with manual activity (Previc, 1998). More generally, executing motor actions requires many of the same operations as intelligence, in that in both cases we must develop hierarchical, sequential organizational strategies for achieving goals – for example, sub-goals must be established that have to be achieved before obtaining the main goal. We must flexibly adjust our course of action depending on obstacles or events in the environment, and we must maintain goals, target positions etc. in working memory. Finally, active involvement using motor circuits is known to improve learning and memory relative to passive learning environments (Hein, 1974).

The link between motor behavior and intelligence is further strengthened by the involvement of dopamine-rich motor association cortex in higher-order mental operations. The area most involved in grammatical comprehension – known as area 45 or Broca's area[10] – may be considered part of orofacial motor association cortex, while the area most involved in mental calculations and working memory – area 46 – may be considered part of oculomotor association cortex (Previc et al., 2005). The link between intelligence and motor behavior is further supported by the greater importance of the dopamine-rich left hemisphere relative to the right one in voluntary motor behavior and motor

[10] This general area was first recognized by the nineteenth-century neurologist Paul Broca as being important to speech and language.

programming (e.g. Greenfield, 1991; Kimura, 1993) as well as in fluid intelligence (see Previc, 1999).

3.2.2   Working memory

As already noted, working memory is arguably the single-most important skill required for general intelligence (Carpenter *et al.*, 1990; Kyllonen and Christal, 1990). Working memory refers to the ability to store, retrieve, and operate on items in memory on a short-term basis (typically, a few seconds). It is difficult to imagine that we could carry out mental operations well if we could not maintain information in one register while at the same time operating and retrieving information from other registers. During decision-making, which activates lateral prefrontal dopamine neurons (McClure *et al.*, 2004; Sevy *et al.*, 2006), we must hold two or more alternatives in memory while weighing our course of action. During language comprehension, we must likewise be able to process the last part of sentences while simultaneously performing high-level interpretation based on clauses encountered several seconds earlier. Not surprisingly, working memory scores are highly predictive of overall language comprehension abilities (Daneman and Merikle, 1996).

Both human and animal experiments point to the important role of the lateral prefrontal cortex in working memory (see Goldman-Rakic, 1998) along with an assortment of areas with strong interconnections to it (e.g. the parietal eye-fields and lateral temporal lobe). Most regions heavily involved in working memory have strong interconnections with this region. These regions are important not only to working memory but also to mathematical and other complex reasoning tasks. All of these brain areas tend to be very rich in dopamine, which has perhaps more than any other neurotransmitter been implicated in working memory (Ellis and Nathan, 2001).

There is also evidence of an overall left-hemispheric bias in working memory, although it is less compelling and more constrained than in the case of motor programming. The left-hemispheric advantage in working memory is greater when semantic processing is required, but it is also present during difficult and/or novel spatial working memory tasks (see Previc, 1999).

3.2.3   Cognitive flexibility

The ability to alter one's cognitive and motor strategy based on new information, much like an executive constantly strategizing about his company's future direction, is intuitively the cognitive skill most closely synonymous with the notion of executive intelligence. To arrive at optimal problem-solving and decision-making, it is necessary but not sufficient to know and process what information has been provided, as the ability to act on information (feedback) that is discrepant with one's current approach or strategy is also critical. This is especially true with language, where ongoing comprehension requires a constant updating of a sentence's context based on the content of previous words and phrases.

The paramount role of the lateral prefrontal cortex in cognitive shifting has been conclusively established on the basis of numerous tests, the most famous of which is the Wisconsin Card-Sorting Test. In this test, cards with different forms, colors and number (e.g. two red squares) are presented, and the patient is required to sort by one category (e.g. color) and then shift to a different one (e.g. form) without explicit instructions. Sorting on the basis of the first category merely becomes "nonrewarded" while sorting on the basis of the second and later category becomes "rewarded." Patients with prefrontal damage, particularly in the left hemisphere, have great difficulty in performing this task even though they may be aware that the rewarded cue has changed (Barcelo *et al.*, 1997; Smith *et al.*, 2004). The same basic result, including the left-hemispheric dominance, holds true for other cognitive shifting tests, such as the trail-making test in which patients must connect a pattern of alternating numbers and letters (e.g. A-1-B-2-C-3, etc.) (Moll *et al.*, 2002).

The lateral prefrontal cortex's role in shifting is linked to its rich dopamine concentration in that dopaminergic neurons are sensitive to reward contingencies and are part of an inhibitory network that can stop ongoing activity and adjust one's responses on the basis of prior context (Braver and Barch, 2002). Moreover, dopamine activity in the dorsal striatum, which is reciprocally connected to the lateral prefrontal system, is highly altered during shifting (Monchi *et al.*, 2006). Lateral-prefrontal dopamine neurons specifically signal a discrepancy in expected versus actual outcome (Rodriguez *et al.*, 2006). Destruction of prefrontal dopamine systems disrupts behavioral switching in animals, and dopamine antagonists do the same in humans (see Koob *et al.*, 1978; Nieoullon, 2002; Simon *et al.*, 1980). One example of this is when one receives a cue pointing in one direction (e.g. leftward) to make an eye movement in the opposite direction (e.g. rightward); without the lateral prefrontal cortex and a normal dopaminergic system, performance in this so-called "anti-saccade" task suffers severely (Pierrot-Deseilligny *et al.*, 2003). Another example of dopaminergically mediated contextual switching is *latent inhibition*, the tendency to ignore cues that have previously been made irrelevant to a task. Schizophrenics and even highly creative individuals with reputedly low prefrontal activity relative to medial subcortical activity attend to previously rewarded and non-rewarded cues similarly and do not show the latent inhibition context effect (see Section 3.2.6 and Section 4.2.8) Similarly, the loss of dopamine in Parkinson's disease and even normal aging may underlie the loss of cognitive flexibility in those populations (Braver and Barch, 2002; Nieoullon, 2002; Previc, 1999).

3.2.4 *Abstract representation*

The ability to create and maintain abstract representations (e.g. in symbol use and concept formation) is present only in the most intellectually advanced animal species and is considered another component of fluid intelligence in humans (Carpenter *et al.*, 1990). To engage in such behavior requires that the symbol and event be divorced from immediate space and time, as the symbol is spatially displaced from its referent and the concept is divorced from a specific place and time. The most abstract space–time concepts are cosmological and existential ones, such as universe, infinity, heaven, and the afterlife. Without abstract symbols (letters, words etc.), advanced language skills would not be possible – indeed, the ability to traverse the past, present, and future as well as to use completely arbitrary symbols are all critical design features of language (Bickerton, 1995; Suddendorf and Corballis, 1997; Wilkins and Wakefield, 1995). Language, therefore, represents one of the most salient examples of off-line human thought that has no immediate impact on survival (Bickerton, 1995; Suddendorf and Corballis, 1997), in contradistinction to the predominantly present-oriented thought of other primates.

I have already noted how dopaminergic systems originally used for interacting in distant "real" space were expropriated during human evolution for use in even more cognitively distant spatial and temporal realms, including abstraction, mental time traveling, and religious activity. One piece of direct evidence for dopamine's involvement in abstract concept formation and abstract reasoning comes from the Raven's Progressive Matrices Test, which requires an understanding of the conceptual relationship among a series of abstract designs. In this test, a sample of up to nine abstract forms, arranged in a logical order with one form missing, is presented; a set of up to eight alternatives is also presented and the person must choose the correct one to replace the blank one. Dopamine levels are positively correlated with performance on the Raven's test (Nagano-Saito *et al.*, 2004), consistent with:

1. the poor performance of Parkinsonian, aging, and other dopamine-deficient populations on this test (Nagano-Saito *et al.*, 2004; Ollat, 1992);
2. the superior performance of persons with the hyperdopaminergic autistic spectrum disorder (Dawson *et al.*, 2007; Hayashi *et al.*, 2008); and
3. the crucial involvement of the dopamine-rich lateral prefrontal cortex (Gray *et al.*, 2003; Prabhakaran *et al.*, 1997), especially in the dopamine-rich left hemisphere (Berman and Weinberger, 1990).

Indeed, many studies using different tests, including deductive logic, have shown that lateral prefrontal involvement in abstract reasoning is lateralized to the dopamine-rich left hemisphere (Deglin and Kinsbourne, 1996; Gray *et al.*, 2003; Prabhakaran *et al.*, 1997; Villardita, 1985), even if the tests are nonverbal in nature. In fact, the distinction between left-hemispheric "abstract/analytical/propositional" and right-hemispheric "concrete/wholistic" thought is one of the most widely accepted and important functional lateralizations in the human brain (Bradshaw and Nettleton, 1981; Nebes, 1974; Ornstein, 1972). By contrast, the earthier, more emotionally astute and practical right hemisphere is more adept at proverb interpretation, judging speaker's intent, inferring other people's thoughts, and detecting lying, as well as at certain 3-D geometrical skills linked to peripersonal operations. What all of these right-hemispheric skills have in common is that they rely on bodily-centered or emotional signals from the body (e.g., proverbs convey subtle emotional messages, judging speaker's intent involves processing facial expressions or the emotional quality/prosody of the voice, and geometrical processing requires the 3-D visualization of objects, which primarily occurs in body space during manipulations).

3.2.5   *Temporal analysis/processing speed*

A key element allowing both computers and humans the capacity for advanced intelligence is a high information processing speed or "baud rate." Colloquial expressions about "being slow" or "quick-thinking" are often used to describe intellectual capabilities. Moreover, there is strong scientific evidence linking general intelligence to the speed of information processing and memory access, using measures of reaction time and event-related brain activity (Bates and Stough, 1998; Fry and Hale, 2000). A high "baud rate" is essential for the rapid delivery and comprehension of spoken language, which involves transient acoustic patterns in the millisecond range. Indeed, many developmental language impairments are believed to be caused by underlying deficits in temporal processing (Farmer and Klein, 1995; Tallal et al., 1993).

Dopaminergic systems are very important to the timing of our short-duration internal "clock." Dopaminergic agonists speed up the clock, whereas dopaminergic antagonists slow it down (Meck, 1996). These effects are consistent with the slower information processing in Parkinson's disease (Sawamoto et al., 2002), which is not causally dependent on the slowed reaction-times in this disorder. The influence of dopamine in increasing processing speed is also evidenced by the faster reaction-times following intake of drugs that increase dopaminergic transmission, such as piribedil (Schuck et al., 2002) and levodopa (Rihet et al., 2002).

The role of dopamine in speeding up cognitive processing is further consistent with the greater involvement of the left hemisphere in rapid temporal processing. For example, the left hemisphere is more involved in rhythmic aspects of language and music, speech, upper-limb movements, and even tactile perception (Bradshaw and Nettleton, 1981). The paramount role of the left hemisphere in the perception and production of both basic (e.g. articulation) and higher-order (e.g. syntactical) aspects of language appears to be highly dependent on its role in rapid sequential processing and programming in the auditory modality (Tallal et al., 1993).

3.2.6   Generativity/creativity

Despite the first five essential skills, our advanced intelligence would nonetheless be greatly limited were it not for the final dopaminergically mediated skill – the ability to generate new solutions and create and express novel ideas and associations. The incredible generativity of human language, symbol use, musical expression, engineering etc. is unquestionably one of the most distinguishing features of the human intellect relative to those of other advanced species (Corballis, 1992). The ability to generate an almost limitless number of unique sentences using only a few dozen phonemes and symbols, known as "generative grammar," is an essential feature of human spoken and written communication (Chomsky, 1988; Corballis, 1992; Hockett, 1960).

As noted earlier, dopamine promotes associations among events and helps to stimulate mental fluency (the ability to generate words, associations etc). One example of this is diminished verbal and semantic fluency found in dopamine-deficiency conditions such as aging, Huntington's disease, Parkinson's disease, and phenylketonuria (see

Previc, 1999), which can be overcome when drugs that boost dopamine activity are provided (Barrett and Eslinger, 2007). Another interesting example of the role of dopamine in creativity comes from the link between creativity and the previously described phenomenon known as "latent inhibition." Like highly dopaminergic schizophrenics, creative individuals also show a markedly reduced tendency to filter out irrelevant stimuli in the latent inhibition paradigm (Carson *et al.*, 2003). Given the link between creativity and reduced latent inhibition caused by excessive ventromedial dopaminergic activity (Weiner, 2003), as well as the generally inhibitory role of the lateral frontal system, the creative drive appears to involve primarily activation of the more "impulsive" ventromedial dopaminergic pathways (see Flaherty, 2005; Reuter *et al.*, 2005). Another link between dopamine and creativity is that, when trying to generate mental images, words, solutions, or otherwise think creatively, our eyes tend to move upward in the direction of distant space (Falcone and Loder, 1984; Previc *et al.*, 2005). In the extreme, the dopaminergic drive to make stimulus associations leads to superstitious behavior (in the case when the associated events are purely random), a heightened sense of relatedness during mystical states (Previc, 2006), and a "loosening of associations" in schizophrenics, who are more likely to make unusual, remote, and even bizarre associations to verbal stimuli. Seeing unusual relationships is also a hallmark of creativity, so it is not surprising that creativity is anecdotally associated with ventromedially driven states like dreaming (Horner, 2006) or that schizophrenia, mania, attention-deficit hyperactivity disorder and other hyperdopaminergic disorders are all associated with greater creativity in affected individuals and immediate family members (Abraham *et al.*, 2006; Goodwin and Jamison, 1990: Chapter 4; Karlsson, 1974; Previc, 1999).

Corballis (1992) argues that the dopamine-rich left hemisphere contributes more to generativity than does the right hemisphere, and this is certainly true in the case of verbal fluency, sign language, and mental imagery (see Previc, 1999). However, visuoconstructive tasks are usually better performed by the right hemisphere, and it is generally recognized that creativity probably involves a complex set of interactions involving both the left and right hemispheres (Bogen and Bogen, 1988; Hoppe, 1988).

3.3   Dopamine and emotion

While both dopaminergic systems are concerned with either initiating or guiding voluntary (goal-directed) behavior in extrapersonal space, the role of dopamine in emotion is more complex and system-specific. As noted earlier, dopamine is tied to the incentive-based pleasurable sexual act and even promotes some "positive" activational states such as grandiosity, elation, and even euphoria (Alcaro *et al.*, 2005; Burgdorf and Panksepp, 2006; Ikemoto, 2007), but there is less convincing evidence that dopaminergic systems are involved in most emotional arousal or interactions and such social tendencies as warmth, playfulness, and empathy. Rather, proximal socio-emotional affiliative interactions and most emotions depend more on serotonin, norepinephrine, opioids, and oxytocin (Depue and Morrone-Strupinsky, 2005; Nelson and Panksepp, 1998; Previc, 2004), which is why boosting serotonin and norepinephrine is the preferred pharmacological option in treating most mood disorders.

Because of its parasympathetic action, elevated lateral dopaminergic activity in particular may actually serve to dampen emotional arousal (Delaveau *et al.*, 2005; Deutch *et al.*, 1990; Finlay and Zigmond, 1997; Mandell, 1980), which is why emotional stress can frequently precipitate hyperdopaminergic clinical states (see Chapter 4). As already noted, dopamine has been associated with extraversion – but more in the case of "agentic" extraversion (characterized by the use of other people to achieve goals) than "affiliative" extraversion (characterized by genuine empathy for other people) (Depue and Morrone-Strupinsky, 2005; Panksepp, 1999). Indeed, the trait of emotional detachment has consistently been associated with high levels of dopamine and certain dopaminergic receptor genes such as DRD2 (Breier *et al.*, 1998; Farde *et al.*, 1997), and the ability to regulate (or even inhibit) one's emotions is a characteristic of extremely goal-oriented individuals, whose traits most typify the lateral dopaminergic personality (see Section 3.4.2). In monkeys and other animals, dopamine agonists decrease social grooming and increase social isolation (Palit *et al.* 1997; Ridley and Baker, 1982; Schlemmer *et al.*, 1980), just as social withdrawal occurs in hyperdopaminergic disorders such as autism and schizophrenia (see Chapter 4). Heightened aggressiveness has also been shown to occur following dopaminergic stimulation (de Almeida *et al.*, 2005; Miczek *et al.*, 2002).

The restricted involvement of dopaminergic systems in emotional arousal is consistent with the social and emotional deficiencies of the dopamine-rich left hemisphere in most humans. When tested in isolation, the left hemisphere has difficulty in understanding the emotional content of communications, in processing the intonations (prosody) of speech that convey its affective component, in judging speakers' intent (in what has been generally termed "theory of mind"),[11] in producing facial expressions, and in accurately gauging the emotion of another individual from facial expressions (Borod *et al.*, 2002; Happe *et al.*, 1999; Heilman and Gilmore, 1998; Kucharska-Pietura *et al.*, 2003; Perry *et al.*, 2001; Sabbagh, 1999; Weintraub and Mesulam, 1983). Other derivative emotional deficits of the left hemisphere include the reduced ability following right-hemispheric lesions to discern the emotionally laden meaning of proverbs (Bryan, 1988) and to determine if someone is lying from his or her vocal and facial behavior. Although a predominance of left-hemispheric activity can frequently lead to mania (Cummings, 1997; Cutting, 1990; Joseph, 1999), the increased activity level in mania is not always accompanied by improved mood. Moreover, superficially improved mood following right-hemispheric damage can also stem from the left hemisphere's lack of awareness (anosagnosia) of the damage to the other hemisphere (Cummings, 1997; Cutting, 1990).

3.4   The dopaminergic personality

As with emotion, the role of dopamine in defining our individual personalities is more complex and controversial than is the dopaminergic contribution to intelligence. Dopamine excess has been associated with numerous types of normal and subclinical personalities, including impulsive (Comings and Blum, 2000), detached (Farde *et al.*, 1997), extroverted (Depue and Collins, 1999), novelty-seeking (Cloninger *et al.*, 1993), schizotypal (Siever, 1994), and telic or serious-minded (Svebak, 1985). While many of these personalities may seem at first glance to be mutually incompatible, the differences among them may be more apparent than real because of the differential involvement of the major dopaminergic systems.

As already discussed, the lateral and ventromedial dopaminergic systems appear to mediate different functions and it seems reasonable to propose that they mediate different personality traits as well (see Table 3.1). Most evidence suggests that the ventromedial dopaminergic

---

[11]   In one well-known example of judging the intent of others using the "theory of mind," the patient is first shown a picture of a girl trying to find a lost dog after losing it in one place and then is shown a picture of the dog in another place after the girl has left the room. Patients with right-hemispheric damage will be more likely to incorrectly predict that the girl who lost the dog will look for it in its new place, because they cannot take on the mental perspective of the girl (who never actually saw the dog placed elsewhere).

Table 3.1 *Features of the two dopaminergic systems.*

Lateral	Ventromedial
Associated with focal-extrapersonal system	Associated with action-extrapersonal system
Future-oriented (event prediction)	Future-oriented (exploration)
Distal goal-oriented (strategic)	Distal goal-oriented (initial drive)
Rational (abstract)	Creative (paranormal experiences)
Focused, controlled (internal locus)	Hyperactive, impulsive
Unemotional	May mediate emotions such as euphoria and aggression

system, largely associated with the action-extrapersonal brain pathways, has an intense motivational drive and connections with the limbic system and is more likely to be involved in addiction, aggression, impulsive and compulsive behavior, sexual activity, and creative and even paranormal thought. By contrast, the lateral prefrontal cortex, associated more with the focal-extrapersonal brain pathways, is involved in planning and strategizing and provides a major inhibitory control over the intense drives of the ventromedial dopaminergic system. The Freudian analogy (Freud, 1927) of the "ego" guiding and even overriding the "id" is an apt description of the relationship of the lateral to ventromedial dopaminergic systems, although the "id" should not be viewed merely in terms of Freud's "pleasure principle" but rather as a set of impulses, primitive drives, and loosened thought processes.[12] In reality, the two systems considerably overlap in their functions and largely work together, and no one individual has a purely lateral or ventromedial personality. But, the relative balance between them may account for the variation in personality traits even among normal individuals with the same overall dopamine levels.

[12] The "ego" should also not be confused with the "self" – whereas both the "ego" and the "id" are outwardly directed and dopaminergically mediated, the self implies an awareness of one's body, which explains why it is more dependent on right-hemispheric functioning (e.g. Keenan et al., 2000). Freud also postulated the existence of a "superego" that is responsible for guiding both the "ego" and "id" toward socially altruistic behavior and may be considered somewhat analogous to one's "conscience." The "superego" is more difficult to pin down neuroanatomically: although certain religious obsessions akin to a hyperactive conscience may be associated with orbitofrontal activity, especially on the left side (see Previc, 2006), the right hemisphere of most individuals appears to be the site of most prosocial and empathetic behavior (see Section 3.3).

3.4.1   *Ventromedial dopaminergic traits*

The ventromedial dopaminergic systems provide us with an intense but unconstrained and even aggressive motivational drive directed toward distant, incentive-laden goals. It is the inability to control the ventromedial drives and thought patterns that underlie most of the hyperdopaminergic mental disorders to be described in Chapter 4. The nucleus accumbens, a dopamine-rich structure with important connections to the limbic system and ventral prefrontal lobe, is the best-studied structure in this regard. Dopaminergic neurons in the medial shell of the accumbens are highly active during goal-directed activities, especially for incentive-laden goals (Alcaro *et al.*, 2005; Berridge and Robinson, 1998; Blackburn *et al.*, 1992; Ikemoto, 2007; Ikemoto and Panksepp, 1999). By contrast, ventromedial regions do not appear to play as much of a role in executive intelligence as does the lateral system (Cummings, 1995).

As noted in Section 3.1.2, opinion among most researchers has shifted from the view that the ventromedial-dopamine system mediates pleasure sensations ("liking") to the view that this system is responsible for motivational drive ("wanting") (Alcaro *et al.*, 2005; Berridge and Robinson, 1998; Blackburn *et al.*, 1992; Ikemoto and Panksepp, 1999). Functional brain imaging in humans has shown that widespread areas of the ventromedial dopaminergic system, including the orbitofrontal cortex and anterior cingulate, are intensely involved in the cravings of substance abusers (Evans *et al.*, 2006; Goldstein and Volkow, 2002). The medial-dopaminergic motivational drive can range from the impulsive (as in attention-deficit disorder and various risk-taking behaviors including gambling, sexual promiscuity, and some drug addictions) to the compulsive (workaholism and obsessive-compulsive rituals) (see Chapter 4), with the latter involving cortical areas to a greater extent. However, it should be stressed that while the medial-dopaminergic system is implicated in most types of psychological addictions, other neurochemical systems such as the opioid and serotonergic ones are also involved – e.g. serotonergic deficits may promote some types of substance abuse (e.g. alcoholism), partly due to release of serotonergic inhibition on the ventromedial dopaminergic system (Johnson, 2004).

The ventromedial dopaminergic system, including the nucleus accumbens and its shell, is also involved in increasing the salience of extrapersonal stimuli and in making associations between them (Blackburn *et al.*, 1992; Kapur, 2003). As noted earlier, ventromedial dopaminergic over-activity is believed to result in loosened associations among distal stimuli and/or internally generated stimuli, as in the loosened associations and bizarre thought patterns of schizophrenics. Dreaming is a particularly vivid example of what happens when the lateral prefrontal system is quieted and the medial systems (including the anterior cingulate and ventromedial frontal lobe) are unleashed (Solms, 2000; Previc, 2006). The extrapersonal themes, bizarre associations, and often-intense repetitive and anxiety-laden drives represent the epitome of irrationality and closely resemble the psychotic state of schizophrenics (Gottesmann, 2002; Hobson, 1996; Previc, 2006). However, the bizarre, chaotic associations created by the unleashing of ventromedial dopaminergic activity can more positively be associated with creative genius, which relies partly on the ability to generate or detect unusual associations among stimuli. Not only are there phenomenological, evolutionary, and familial links between genius and madness (Horrobin, 1998; Karlsson, 1974), but a link between genius and madness clearly exists in the countless number of famous individuals who have experienced both (Karlsson, 1974), including the Nobel prize-winning author John Nash, who suffered from paranoid schizophrenia. The following conversation between Nash and a colleague Mackey is particularly revealing in this regard:

"How could you", began [Harvard professor George] Mackey, "how could you, a mathematician, a man devoted to reason and logical proof ... how could you believe that extraterrestrials are sending you messages? How could you believe that you are being recruited by aliens from outer space to save the world? How could you ...?" ... "Because" Nash said slowly in his soft, reasonable southern drawl, as if talking to himself, "the ideas I had about supernatural beings came to me the same way that my mathematical ideas did." (Nasar, 1998: 11)

3.4.2   Lateral-dopaminergic traits

The lateral prefrontal dopaminergic system is crucial to maintaining control over behavior and thought. In animals, it has been shown that dopamine levels are very high when stressors are controllable but less so when they are uncontrollable (Anisman and Zacharko, 1986; Coco and Weiss, 2005). When animals are required to actively control their environment by making cost–benefit decisions about rewards – e.g. choosing a larger but delayed reward over a smaller but immediate one – lateral prefrontal dopaminergic systems are also stimulated (Denk et al., 2005; McClure et al., 2004; Sevy et al., 2006). These findings are generally consistent with dopamine's role in mediating active exploration, achievement motivation, and goal-seeking, particularly in males

(Previc, 2004). The lateral dopaminergic system – originating from the focal-extrapersonal system involved in oculomotor search, scanning, and memory – essentially channels the less-constrained medial-dopaminergic systems into organized plans and motor sequences to achieve the intended motivational goal. It is the lateral dopaminergic system that is most closely tied to the executive skills reviewed in Section 3.2 and is presumably the basis of the "telic" (intellectual and serious-minded) personality described by Svebak (1985) and the "agentic" extrovert personality (Depue and Collins, 1999).

As noted previously, lateral dopaminergic systems may also be crucial in maintaining what has been termed an *internal locus-of-control* (see review by Declerck *et al.*, 2006). Locus-of-control refers to the tendency to view one's choices and experiences as being under one's own control (known as "internal" control) or as determined by fate or others (known as "external" control). Having a high internal locus-of-control is believed to be an important predictor of success in life and in combating stress and disease (De Brabander and Declerck, 2004; Kushner *et al.*, 1993; Regehr *et al.*, 2000) and in generally surviving extreme environments (Previc, 2004). It is also greater in males (De Brabander and Boone, 1990), is associated with left-hemispheric function (De Brabander *et al.*, 1992), and is correlated with high goal-directedness, executive intelligence, achievement motivation and a future (as opposed to present) temporal perspective (Declerck *et al.*, 2006; Markus and Nurius, 1986; Murrell and Mingrone, 1994). High dopamine levels support an internal locus-of-control belief (Declerck *et al.*, 2006), as exemplified by the effect of dopamine-boosting drugs such as amphetamine in the lateral prefrontal cortex to increase perceptions of internal control in those suffering attention-deficit disorder and certain medical diseases (Pelham *et al.*, 1992). Because of its ability to inhibit the limbic stress response (Deutch *et al.*, 1990; Finlay and Zigmond, 1997) and to regulate or even inhibit emotionality (Declerck *et al.*, 2006), the lateral prefrontal dopaminergic system is especially useful for clear thinking under stress. Accordingly, lateral-dopaminergic traits frequently predominate in successful military leaders, who because of their intelligence and planning skills, belief in their ability to control events, ability to function under stress, and reduced emotional attachments, may be especially well-suited to exert bold leadership in dangerous situations (Previc, 2004).

The two greatest weaknesses of the lateral prefrontal system are:

1. over-focusing/excessive inhibition, traits that are found in persons with autism (see Chapter 4); and
2. an exaggerated belief in one's own power to control people and events.

Too much dopamine can propel one's internal control sense to abnormal limits, as illustrated by the delusions of grandiosity following administration of amphetamine (Krystal *et al.*, 2005), which is the preferred treatment to combat under-focusing and lack of control in attention-deficit disorder. Extreme dopamine activation may lead to a belief of individuals that they can control thoughts and events at great distances and are in the midst of cataclysmic and cosmic forces acting through them, as in mania and schizophrenia (see Chapter 4), but even mild dopamine elevations can lead to schizotypy and magical ideation, in which individuals may feel that they have special powers and/or the ability to control random events (Previc, 2006).

As already noted, most individuals do not exhibit exclusively one set of dopaminergic personality traits. As will be highlighted in Chapter 6, most great historical figures have possessed both lateral and ventromedial traits – i.e. a high dopaminergic intelligence and strategic (distal or far-sighted) vision along with creative, impulsive, mystical, and even irrational dopaminergic tendencies.

3.4.3   Dopamine and the left-hemispheric (masculine) style

As noted throughout this chapter, dopaminergic traits such as distal orientation, abstract intelligence, paranormal beliefs, internal locus-of-control, and reduced emotionality characterize the dopamine-rich left hemisphere of most humans, as determined by unilateral stimulation of it, isolated testing of it after severance of the connections between it and the right hemisphere (i.e. the split-brain patient),[13] and following damage to the right hemisphere. General descriptions of the left hemisphere in most individuals as more analytical, goal-oriented, linear-thinking, and controlled are widely accepted (Bradshaw and Nettleton, 1981; Nebes, 1974; Ornstein, 1972; Tucker and Williamson, 1984). The left hemisphere also appears to be more involved in aggression (Andrew, 1981; Elliott, 1982; Mychack *et al.*, 2001; Pillmann *et al.*, 1999) and sexual behavior (Braun *et al.*, 2003). Hyperdopaminergic disorders such as autism, mania, obsessive-compulsive disorder, and schizophrenia are all more likely to occur following right-hemispheric than left-hemispheric damage and thereby reflect left-hemispheric overactivation (see Chapter 4). With its extrapersonal drives, the left hemisphere is not only more

---

[13] The split-brain patient has his or her corpus callosum connecting the right and left hemispheres severed, which is performed in rare cases to alleviate otherwise treatment-resistant epilepsy and allows researchers using appropriate tests to assess the function of the isolated left or right hemisphere.

likely to house our dreams and religious experiences (Previc, 2006; Solms, 2000) but also the hypothesis-testing scientist within us (Wolford et al., 2000). Conversely, as already reviewed, the right hemisphere – richer in norepinephrine and serotonin – is less involved in extrapersonal space, analytical intelligence, sexual behavior, and aggression but is more inclined toward peripersonal activity, self-awareness, and social empathy.

It is worth noting that the active, analytical, controlled, less emotional, sexually driven, and even aggressive style of the left hemisphere is more typical of male behavior in general, and it is popular to view male behavior as predominantly "left-brained" (Gray, 1992; Ornstein, 1972). Indeed, the left- and right-hemispheric modes are similar to the ancient Chinese concepts of "yang" (masculine, upward-seeking, and active) versus "yin" (feminine, downward-seeking, and passive), which, in turn, have their parallels in many other myths and concepts (Ornstein, 1972). While most males and females alike both possess left-hemispheric dominance for language and other functions, the left-hemisphere's dopamine-related active, controlling, and even aggressive style suggests a greater "masculinity," whereas the right-hemisphere's greater emotional sensitivity and empathy mediated by its relatively greater noradrenergic and serotonergic concentrations suggests a greater "femininity." As previously noted, testosterone increases overall dopaminergic activity and, as will be reviewed in Chapter 4, males are more prone to almost every hyperdopaminergic disorder.[14] By contrast, estrogen inhibits dopamine activity and promotes a more passive, receptive style – for example, classic female sexual receptive postures such as lordosis are stimulated by norepinephrine but inhibited by dopamine (Robbins and Everitt, 1982). Not surprisingly, most studies have shown that males around the world are generally more likely to adopt internal locus-of-control and futuristic perspectives (Bentley, 1983; de Brabander and Boone, 1990; Greene and Wheatley, 1992; Sundberg et al., 1983). The

---

[14] Geschwind and Galaburda (1985) were among the first to emphasize the possible relationship between the left hemisphere and male-dominated disorders. They argued that testosterone had a direct effect on delaying the development of the left hemisphere, which is incorrect because it is damage to the right hemisphere that mimics most of the male-biased psychological disorders (see Chapter 4). In reality, the typical male brain is more characteristic of the cognitive and emotional style of the left hemisphere, not necessarily because the male brain consists of an overactive left hemisphere but because the dopamine content in *both* of its hemispheres is exaggerated. It must be stressed, however, that the male bias in dopamine function is highly variable and hardly immutable. Indeed, as females achieve greater success in modern societies, the dopaminergic content of their brains may be increasing, which could be a factor in the rising incidence of autism (Previc, 2007).

particular relationship between masculinity and the left hemisphere may have been of great historical significance in the rise of male-dominated dopaminergic societies at the end of the Neolithic period, as discussed further in Chapter 6.

The combined role of dopamine in reward prediction, stimulus associations, motivation, and control may help to understand the intriguing finding of Wolford *et al.* (2000), whose paradigm tested the responses of the isolated left and right hemispheres in a split-brain patient to cues that randomly signaled future reward at a higher or lower probability, depending on whether the lights were on the left or right side. Using only the dopamine-rich left hemisphere, Wolford *et al.*'s patients futilely attempted to derive an underlying pattern and ended up doing poorly. This strange left-hemispheric behavior reflects its intense need to predict and control future rewards along with a desire to find *abstract* patterns or relationships – all of which are manifestations of an orientation beyond immediate space and time. By contrast, the same patients using only their earthier right hemisphere used a simple strategy and merely pressed the more highly rewarded key, thereby performing much better in the end. In fact, the behavior of the isolated right hemisphere was more similar to that of birds and nonhuman mammals, whose overall brain dopamine content is much lower than that of humans. This need to predict and control external events paradoxically underlies scientific knowledge and superstitious behavior, both of which are products of the left hemisphere's highly dopaminergic mind (Previc, 2006).

3.5   Summary

Based on the preceding review, it can be concluded that every major behavior or trait associated with dopamine can either directly or indirectly be tied to a more primordial link between dopamine and distant space and time. Based on the neuroscientific literature, it may be predicted that highly dopaminergic minds are:

1. fairly active in their intellectual and possibly physical lifestyle;
2. above-average in intelligence, with a particularly impressive working memory and strategic ability;
3. very achievement- (goal-) oriented, almost to the point of obsessiveness;
4. always seeking new goals, especially incentive-laden ones such as money, fame, power, or idealistic achievements, yet becoming restless once those goals are achieved;

5. very confident in their ability to control their destiny and dominate over other people, sometimes to the point of grandiosity and recklessness;
6. more aggressive than compassionate or nurturing toward other people; and
7. above-average in sexual desire, especially if male, but not necessarily overly hedonistic in other respects.

If a person's high dopaminergic content were biased toward the lateral prefrontal dopaminergic systems, he or she would tend to be more analytic, controlled, and highly fixated on objects or ideas. If that person's high dopaminergic content were concentrated in the ventromedial dopaminergic system, and especially its subcortical elements, he or she would tend to be more creative, active, animated, restless and, in the extreme, aggressive and delusional.

Do you know any such individuals with high dopamine contents? Do you know societies that are dominated by such individuals? If not, I will introduce them to you in Chapter 6, because their influence on the course of human history has been profound. Not only did many of these individuals produce huge discoveries, conquests, and achievements, but they almost invariably suffered from a dark side that wreaked havoc on close family members or, in some cases, entire populations. The other side of the dopaminergic mind will be discussed at length in the next chapter, which will highlight the most prominent of the various hyper-dopaminergic clinical disorders.

4   Dopamine and mental health

4.1   The "hyperdopaminergic" syndrome

Despite the many positive dopaminergic traits described in Chapter 3, the dopaminergic story has another, darker side. Whereas too little dopaminergic transmission in disorders such as Parkinson's disease and phenylketonuria is debilitating to motor and intellectual functioning, excessive dopamine activity in one or more brain systems has been implicated in an even larger number of prominent neuropsychological disorders – including attention-deficit disorder (also known as attention-deficit/hyperactivity disorder when accompanied by hyperactivity), autism, Huntington's disease, mania (also known as hypomania and bipolar disorder when it alternates with depression), obsessive-compulsive disorder, schizophrenia, and Tourette's syndrome. Other hyperdopaminergic disorders include substance abuse (highly associated with attention-deficit/hyperactivity disorder and bipolar disorder) and stuttering (linked to Tourette's syndrome). All of the hyperdopaminergic disorders are closely related to one another in terms of their co-morbidities and symptoms, and more than one set of hyperdopaminergic symptoms are surprisingly often found in the same individual (e.g. autism with obsessive-compulsive and Tourette's features; mania with schizophrenic-like psychosis and obsessive-compulsive behavior) or within families. The various hyperdopaminergic disorders also are highly amenable to the same pharmacological interventions (principally D_2 receptor-blocking drugs). In fact, drugs that decrease dopamine levels in either the brain generally or in a specific system (e.g. the ventromedial one) are the main or secondary pharmacological treatment of choice in every hyperdopaminergic disorder except for Huntington's desease.

Dopamine excess contributes to the motor symptoms (e.g. hyperactivity, tics, motor stereotypies), delusions and hallucinations, and social withdrawal found to varying degrees in the above disorders. Like the hypodopaminergic disorders, the hyperdopaminergic disorders also affect intellectual performance in a mostly negative way. However, it has

75

already been noted how disorders such as schizophrenia, mania, and less severe types of autism are often associated with intellectual genius in those afflicted, how milder versions of these disorders are even more associated with creative genius, and how superior intellects are more frequently found among first-degree unaffected relatives (Karlsson, 1974). As reviewed in the individual sections to follow, most of the hyperdopaminergic disorders have risen in prevalence during the past few decades and now arguably constitute collectively the greatest threat to mental health in the industrialized world (8–10 percent prevalence for attention-deficit/hyperactivity disorder; ~1–3 percent each for bipolar disorder, obsessive-compulsive disorder, schizophrenia, and Tourette's syndrome; and >0.5 percent for autism).[1] Finally, a male excess is found at least in the early-onset and usually most severe form of these disorders, consistent with the link between testosterone and dopamine.

There are several models, embracing both genetic and nongenetic influences, of how dopamine can deleteriously affect mental health. Genetic influences have long been suspected to play an important role in the hyperdopaminergic disorders, based on the greater concordance rates for monozygotic (identical) twins relative to dizygotic (fraternal) twins and other siblings. (Concordance refers to the percentage of twin pairs sharing a disorder – e.g. a 60 percent concordance rate would mean that there is a 60 percent likelihood that one member of a twin pair has a particular trait if the other one does.) It is widely believed that a higher concordance rate for identical twins (which have the same genetic makeup) than for same-sexed fraternal twins (which derive from separate embryos and are no different from regular siblings in their genetic material held in common) implies a genetic influence in a given disorder. However, the fact that two-thirds of identical twins – but no dizygotic twins – also share the same chorion (placental blood supply) is a major problem for genetic estimates based on twin studies (see Prescott et al., 1999). Monochorionic twins have been shown, to varying degrees, to be more similar than dichorionic twins on a host of behavioral and physiological measures, including intelligence, birthweight, and risk for psychopathology (Davis et al., 1995; Melnick et al., 1978; Scherer, 2001). The effects of prenatal exposure to drugs, stress, and infection are all highly influenced by placental type (Gottlieb and Manchester, 1986; Sakai et al., 1991), and this chorionic-genetic

---

[1] Although major depression is the single largest psychological disorder in the United States with a prevalence of ~15 percent, the combined lifetime prevalence of the hyperdopaminergic disorders approaches that figure and the chronic costs may be much greater. For example, adults with autism suffer an umemployment rate of 70 percent and a mean annual income for those employed of less than $4,000 (Bellini and Pratt, 2003).

masquerade appears to be greatest when prenatal effects are strongest (Prescott *et al.*, 1999), as in disorders such as autism and schizophrenia (Davis *et al.*, 1995). Moreover, identical twins are much more similarly reared by parents than are fraternal twins (Mandler, 2001). Nevertheless, genetic heritability estimates from twin studies greater than 50 percent, even if inflated, point to at least some genetic influence in a particular mental disorder. For example, it has been shown that alterations to key dopamine genes such as dopamine-beta-hydroxylase – which converts dopamine to norepinephrine and whose deletion results in too much dopamine and too little norepinephrine – creates a greater likelihood of acquiring attention-deficit/hyperactivity disorder, Tourette's syndrome, and other dopamine-related disorders (Comings *et al.*, 1996). Other clinical studies have shown effects due to genetic disruption of dopamine receptor genes (e.g. polymorphisms) and dopamine transport genes.

However, there are well-documented prenatal and perinatal disturbances that also affect the risk for one or more of these disorders, including maternal drug use (autism, attention-deficit/hyperactivity disorder, bipolar disorder), maternal fever (autism, schizophrenia), and hypoxia at birth (attention-deficit/hyperactivity disorder, autism, schizophrenia). As discussed in Chapter 2, these effects in most cases are associated with elevated dopamine: e.g., immune reactions and associated fever require dopamine release to decrease temperature; known teratogenic agents like thalidomide, various stimulants, and anti-seizure medications increase dopaminergic transmission; and hypoxia stimulates dopaminergically mediated parasympathetic mechanisms to reduce oxygen consumption. Another example of the contribution of prenatal factors to the hyperdopaminergic disorders is deletion of the dopamine beta-hydroxylase gene in mothers, which increases dopamine relative to norepinephrine in the placental blood supply and increases the risk of autism in offspring even more than does the absence of that same gene in the offspring themselves (Robinson *et al.*, 2001). Nongenetic/prenatal factors that contribute to dopamine elevation – including indirect ones such as demographic status and societal pressures (see later discussion) – are more suspect in disorders that recently have been on the rise, such as attention-deficit/hyperactivity disorder, autism, bipolar disorder (mania), and possibly obsessive-compulsive disorder and Tourette's syndrome, since our genetic makeup has not substantially changed in the past few decades. It is especially difficult for genetic factors to explain the stable or rising incidences of disorders such as autism and schizophrenia, since afflicted individuals are unlikely to marry and pass on their genes because of their social inadequacies (see Shapiro and Hertzig, 1991).

In contrast to drugs that increase dopamine, drugs that increase serotonin and norepinephrine (the right-hemispheric neurotransmitters involved in emotion and peripersonal sensory processing) not only improve mood and relieve anxiety (Nelson *et al.*, 2005) but are therapeutically beneficial against the hyperdopaminergic disorders. Conversely, individuals with chronically low serotonin and norepinephrine levels associated with an underlying trait anxiety (see Tucker and Williamson, 1984), or transient depletion of serotonin and noradrenaline due to sleep deprivation and other psychological stressors (see Previc, 2004), are more prone to develop hyperdopaminergic psychopathologies such as schizophrenia. It is important in this context to understand that both norepinephrine and serotonin – but especially the latter – have reciprocal, inhibitory interactions with dopamine, such that greater dopamine concentrations in the brain may produce less noradrenaline and serotonin, and vice versa. Indeed, the inhibitory action of serotonergic systems over dopaminergic ones is extremely well-documented (see Damsa *et al.*, 2004; Previc, 2006, 2007) and is arguably the most significant neurochemical interaction in the entire brain from the standpoint of clinical neuropsychology. In concert with the notion of hyperdopaminergic disorders, there is also a "serotonergic dysfunction disorder" stemming from reduced serotonergic function (Petty *et al.*, 1996). Although these two categories are not identical, there is a strong overlap between them. In fact, the use of dopamine-blocking drugs, serotonin-boosting drugs, or their combination is the preferred treatment in almost every major psychological disorder (e.g. Petty *et al.*, 1996).

The previously described rise of dopamine during stress is, up to a certain point, beneficial. In terms of physiology, dopamine helps to dampen the arousal response by activating parasympathetic circuits, which tend to reduce heart rate, blood pressure, and oxygen consumption (see Chapter 2). In terms of behavior, dopamine helps us by promoting active coping with stress – whether that means stimulating escape behavior in a rat exposed to intermittent shock (Anisman and Zacharko, 1986), in performing problem-solving on a battlefield (Previc, 2004), or in merely helping us to cope with an uncertain socioeconomic environment in which layoffs, divorces, emotional separations etc. are extremely common. Indeed, dopamine may often help to transform our underlying stress, such as a creative person's internal tension, into an intense motivational drive required to achieve a goal and in so doing at least temporarily reduce the anxiety. Elevated dopamine levels, along with norepinephrine and serotonin, may in their stress-dampening roles be crucial ingredients of what is known as the "hardy" or "tough" personality (see Previc, 2004). Dopamine, in particular, instills a belief in individuals that they, not fate, are in control of their destiny – i.e. the internal locus-of-control trait described in Chapter 3 (Declerck *et al.*, 2006). If carried too far, however, active coping, high achievement motivation, excessive internal locus-of-control beliefs, and other such traits associated with dopaminergic over-activation can be extremely debilitating, both in the coping individual (e.g. delusions, social detachment, etc.) as well as in the mental health of their offspring. Autism, in particular, is a hyperdopaminergic disorder that is much more likely to be found in offspring of highly successful parents (Previc, 2007).

In the remainder of this chapter, I will briefly review the etiology and neural basis of the major dopaminergic disorders, with special reference to their pharmacological imbalances, genetic/environmental influences, developmental time-courses, and possible hemispheric asymmetries. Two hypodopaminergic disorders (Parkinson's disease and phenyl-ketonuria) will be reviewed, along with six clearly hyperdopaminergic disorders (autism, Huntington's disease, obsessive-compulsive disorder, mania, schizophrenia, Tourette's syndrome) and another disorder – attention-deficit/hyperactivity disorder – that may involve a selective overactivation of the ventromedial dopaminergic system in conjunction with reduced lateral prefrontal dopaminergic activity. Superficial dis-similarities among the hyperdopaminergic disorders will be accounted for on the basis of different genetic and developmental influences (e.g. autism reflects both genetic and early to mid-gestational influences, schizophrenia is influenced by genetic and mid-to-late prenatal effects, Huntington's disease is the most genetically determined etc.) and by the brain region affected. Furthermore, subcortical dopaminergic systems are more affected in some disorders (e.g. autism, Tourette's syndrome), whereas varying degrees of disturbance to specific cortical dopaminergic systems may be present in other disorders (e.g. mania).

4.2   Disorders involving primary dopamine dysfunction

4.2.1   *Attention-deficit/hyperactivity disorder*

Attention-deficit with hyperactivity disorder is the most prevalent of the learning disabilities, afflicting at least 8 percent of all children in the United States, with a 2.5-fold greater prevalence in males than females (Biederman and Faraone, 2005; Centers for Disease Control and Prevention, 2005). This disorder is characterized by boredom, distractibility, and impulsivity and is frequently associated with childhood depression (Brumback, 1988). According to Papolas and Papolas (2002), attention-deficit/hyperactivity disorder is likely to be co-morbid in the majority of those diagnosed with bipolar disorder, whereas the rarer bipolar diagnosis can be applied to about 30 percent of those with attention-deficit/hyperactivity disorder (Geller et al., 2004). Attention-deficit/hyperactivity disorder is also highly associated (10–30 percent co-morbidity) with autism, obsessive-compulsive disorder, substance abuse, and Tourette's syndrome (Geller et al., 2004; Gillberg and Billstedt, 2000; Kalbag and Levin, 2005; Stahlberg et al., 2004) and, to a lesser extent, with schizophrenia/psychosis (Geller et al., 2004). Attention-deficit/hyperactivity disorder is clearly on the rise, having increased almost three-fold between 1991 and 1998 alone (Robison et al., 2002).

The consensus of researchers is that attention-deficit/hyperactivity disorder is primarily caused by alternations in brain norepinephrine and dopamine (Biederman and Faraone, 2006; Pliszka, 2005), although it has long been an issue as to whether dopamine is elevated or deficient in this disorder (see Pliszka, 2005). In favor of the elevated dopamine hypothesis, attention-deficit/hyperactivity disorder has many features in common with highly co-morbid disorders such as autism, obsessive-compulsive disorder, and Tourette's syndrome that are more definitively linked to excessive dopamine. Also, elevated dopamine in animals either leads to or is associated with hyperactivity, especially in animal models such as the spontaneous hypertensive rat and the Naples High Excit-ability rat (Pliszka, 2005; Viggiano et al., 2003; see Chapter 3). On the other hand, stimulants such as methylphenidate – a drug similar to but somewhat milder than amphetamine – that increase dopamine along with norepinephine are the currently preferred treatment for attention-deficit/hyperactivity disorder, at least in childhood. One leading hypothesis is that the ventromedial dopamine systems are overactive in this disorder whereas the lateral prefrontal pathways that provide inhibitory, executive control to focus our intellectual drive are under-active (Viggiano et al., 2003). This hypothesis is consistent with evidence of dysregulated frontal-striatal circuits in attention-deficit/hyperactivity disorder (Biederman and Faraone, 2005) and with the high co-morbidities of this disorder with obsessive-compulsive disorder, substance abuse and various impulse-control behavioral states in which the ventromedial dopaminergic system is particularly active (Galvan et al., 2007). It may also help to explain why, as the prefrontal associ-ation regions fully mature in adulthood, attention-deficit/hyperactivity disorder begins to wane (Faraone et al., 2006).

Attention-deficit/hyperactivity disorder is believed to have a mostly genetic etiology, with heritability coefficients ranging up 80 percent (Biederman and Faraone, 2005). Most research to date has focused on various dopamine genes, including the D_2, D_4, and D_5 receptor genes, the dopamine transport gene, and the dopamine-beta-hydroxylase gene (Biederman and Faraone, 2005; Comings et al., 1996; Li et al., 2006). However, attention-deficit/hyperactivity disorder has also been linked to a variety of prenatal and perinatal influences that are believed to elevate offspring dopamine levels, including maternal smoking, maternal psychosocial stress, and hypoxia at birth (Biederman and Faraone, 2005). A particularly revealing aspect of attention-deficit/hyperactivity disorder is its well-documented association with right-hemispheric dysfunction, particularly in the context of right-hemispheric deficits and childhood depression (e.g. Brumback, 1988; Heilman et al., 1991). This relationship is consistent with a variety of left-sided tactile and visuo-spatial deficits (known as "soft" neurological deficits) that mimic the attentional disturbances produced by actual damage to the right hemisphere. Right-hemispheric dysfunction would, of course, be expected to shift the balance of hemispheric activity toward the left hemisphere and its already predominant dopaminergic mode.

4.2.2   Autism

Autism and related disorders such as Asperger's syndrome represent the fastest growing of all neurodevelopmental disorders (Previc, 2007). They are characterized by impaired social and emotional relationships and by stereotyped behaviors that take the form of rocking, whirling, head-banging etc. in extreme cases, and obsessive verbal behavior (such as repetitively focusing on trivia) in higher-functioning cases. These disorders are usually apparent by two to three years of age and often earlier. Autism exhibits the most extreme male bias (4:1) of all the major neuropsychological/neurodevelopmental disorders, and this bias is still greater when autism is accompanied by otherwise normal intellectual functioning. Autism was reported in <0.05 percent of the population in industrialized countries as little as three decades ago, but its growth rate has been exponential and recent surveys now estimate its prevalence at slightly greater than 0.5 percent (Previc, 2007).[2] In addition to its high

[2] According to the Centers for Disease Control and Prevention, autism prevalence in elementary-age children in the United States is as high as one in 166: www.cdc.gov/od/oc/media/transcripts/t060504.htm.

co-morbidity with attention-deficit/hyperactivity disorder, autism is also likely to be associated with Tourette's syndrome in about 10–20 percent of high-functioning cases (Gillberg and Billstedt, 2000) and may even more frequently present with obsessive-compulsive symptoms. In fact, there are as many obsessions and compulsions in high-functioning autistics as there are in obsessive-compulsive patients themselves (Russell *et al.*, 2005), leading some researchers to argue for a combined "obsessive-compulsive/autistic" syndrome (Gross-Isseroff *et al.*, 2001). Autism is also co-morbid with bipolar disorder and schizophrenia in slightly less than 10 percent of cases (Stahlberg *et al.*, 2004).

As reviewed by Previc (2007), the etiology and neural substrate of autism remain somewhat unclear, although a great deal of evidence is accumulating that excessive dopamine is a major correlate of autism. There have been a large number of brain areas implicated in autism, but the only consistent findings to date point to various subcortical abnormalities involving the brainstem and possibly the cerebellum and basal ganglia (Miller *et al.*, 2005; Nayate *et al.*, 2005). The involvement of the basal ganglia partly explains the over-focusing and attention to visual details, which along with an impressive fluid intelligence (Dawson *et al.*, 2007; Hayashi *et al.*, 2008) is consistent with an overactivation of the lateral dopaminergic (focal-extrapersonal) pathways in autism. Most of the affected regions – and the brainstem in particular – develop early in gestation, which is consonant with the predominantly first-trimester influence of prenatal teratogenic substances that increase the risk of autism, such as the sleep agent thalidomide and the anti-seizure medication valproic acid (Miller *et al.*, 2005; Stromland *et al.*, 1994). Prenatal exposure to valproic acid in particular is known to increase frontal cortical dopamine levels in a rodent model for autism (Nakasato *et al.*, 2008). In terms of neurochemistry, most research into the causes of autism has focused on dopamine and serotonin. There is little doubt that stereotypical behaviors in autism are the result of dopaminergic over-activation (Ridley and Baker, 1982) and that drugs such as risperidone that block dopamine action are the most widely used and effective against autistic symptoms (Nagaraj *et al.*, 2006; Volkmar, 2001), whereas drugs such as amphetamine that increase dopamine levels exacerbate autistic behavior (Volkmar, 2001). Although many persons with autism have higher blood levels of serotonin, serotonin levels in the autistic brain actually tend to be reduced relative to normals (Previc, 2007). Excessive dopamine is known to impair social behavior in monkeys (Palit *et al.*, 1997; Schlemmer *et al.*, 1980), whereas serotonin reduces social anxiety and promotes social interaction (e.g. Young and Leyton, 2002). Moreover, an imbalance of dopaminergic over serotonergic activity is suggested by subtle disturbances in the concept of personal space in autistics (Parsons *et al.*, 2004). It is not clear whether the emotional and social deficits in autism are caused primarily by the dopaminergic or serotonergic abnormalities, but even the latter may be expressed indirectly through dopaminergic overactivation because of the previously described reciprocal inhibitory interactions involving dopamine and serotonin.

Although autism is reputedly one of the most genetic of all neuropsychological disorders, its risk has also been shown to be increased by a wide range of transient prenatal and perinatal factors that elevate dopamine, including maternal stress, illness, and drug use during pregnancy and hypoxia and cesarean delivery at birth (Previc, 2007). Chronic influences such as maternal age, intelligence, personality, work stress, coping styles, etc. may exert an even greater impact on prenatal dopamine levels. Autistic children are four times more likely to be born to highly educated mothers than to poorly educated ones (Croen *et al.*, 2002), because the former presumably have higher dopamine levels associated with high achievement motivation levels and a greater ability to cope with the uncertainties and pressures of modern societies (Previc, 2007). Some genetic disorders in which autism is especially prevalent (e.g. tuberous sclerosis) are linked to genes that control dopaminergic transmission, but it may be the mother's genes via their indirect effects on prenatal transmission that exert the greatest genetic influence over the offspring's dopamine levels. For example, the risk for autism is doubled when there is a complete absence in mothers of the dopamine beta-hydroxylase gene (Robinson *et al.*, 2001).

One final piece of evidence suggesting that an imbalance between dopamine and serotonin is critical to the etiology of autism is that autistic symptoms generally resemble those following right-hemispheric brain damage (McKelvey *et al*, 1995). Whereas the serotonin-rich right hemisphere is much better at emotional processing, empathy, social interactions, "theory of mind," and processing global concepts rather than local details, the isolated dopamine-rich left hemisphere behaves much more like that of a mildly autistic person of normal or above-average intelligence (Ozonoff and Miller, 1996; Previc, 2007; Sabbagh, 1999). However, right-hemispheric damage alone cannot as easily simulate the extreme, presumably subcortically mediated stereotypical behaviors found in severe autism.

4.2.3   Huntington's disease

Huntington's disease is one of only a few neurological disorders traceable to a single gene, in this case the HD gene (see Purdon *et al.*, 1996). This disease was first described by George Huntington in 1872 and in North America is primarily found in descendants of a small group of Anglicans who immigrated to Long Island, New York, in 1642. Huntington's disease is characterized by involuntary movements of the upper limbs and facial musculature, known as choreic or dancing movements (hence, the frequently used term Huntington's chorea). Huntington's disease is much rarer than the other hyperdopaminergic disorders reviewed here, with a prevalence of only five per 100,000, with most cases developing between thirty and forty-five years of age. It is not primarily a neurochemical disorder per se, as it involves neurodegeneration in the corpus striatum, beginning with the caudate nucleus and later extending into the putamen, globus pallidus, and subthalamic nucleus. Most of the striatal degeneration occurs in small neurons with spiny dendrites, which rely mostly on gamma aminobutyric acid (GABA) transmission. While dopaminergic neurons and receptors (principally D_1 and D_2) are also damaged (Bäckman and Farde, 2001), the net result of disabling GABA transmission appears to be an overall increase in dopaminergic activity (Klawans, 1987). The choreic movements and subsequent psychotic symptoms associated with Huntington's disease can be mimicked by dopaminergic drugs such as L-dopa and reduced by dopaminergic antagonists as well as drugs that increase GABA and acetylcholine, both of which inhibit dopamine in the striatum (Klawans, 1987; Purdon et al., 1996). It is less clear, however, whether cognitive symptoms in Huntington's disease reflect excessive dopaminergic activity or hypodopaminergia (Bäckman and Farde, 2001), as both imbalances can disrupt cognitive performance (Bäckman et al., 2006).

Other choreas such as Sydenham's chorea are nongenetic in origin and are brought about by fevers and other indirect catalysts of increased dopaminergic activity. In general, choreic symptoms overlap hyperdopaminergic motor symptoms found in attention-deficit/hyperactivity disorder, obsessive-compulsive disorder, and Tourette's syndrome (Maia et al., 2005), especially in the early-onset phase. However, there is no evidence for a male predominance in the choreiform disorders, as in the other hyperdopaminergic disorders.

4.2.4   Mania (bipolar disorder)

Mania and its milder version (hypomania) are characterized by heightened verbal, motor, sexual, intellectual, and other goal-directed behavior. Mania is referred to as bipolar disorder when it fluctuates with depression. Cycling between mania and depression is known as Bipolar I disorder, cycling between less extreme hypomania and depression is known as Bipolar II disorder, and long-term cycling of even less severe mood states is known as cyclothymia. Mania is considered part of seasonal affective disorder when it is triggered more in the summer months (when temperature and light levels are higher in moderate-to-extreme latitudes) and depression is triggered more in the winter months (when temperature and light levels are lower in those same latitudes). In addition to elevated motor activity, mania is also characterized by heightened cognitive activity ranging from increased creativity to delusions and other psychotic features (Goodwin and Jamison, 1990). Indeed, of all of the major mental disorders, mania and hypomania are most associated with heightened creativity and intelligence (Goodwin and Jamison, 1990), and bipolar disorder is suspected in many famous persons in history, as will be discussed in Chapter 6. Mania and schizophrenia are also the two disorders most convincingly associated with dopamine-mediated hyper-religiosity (Previc, 2006).

Bipolar I and II disorder together conservatively afflict 1–2 percent of the adult population (Bauer and Pfennig, 2005; Berrettini, 2000; Narrow *et al.*, 2002), although estimates range as high as 6.5 percent with less restrictive diagnostic criteria. Although no overall gender differences exist in the prevalence of bipolar disorder, it is more likely to present as mania in males, especially before twenty-five years of age (Arnold, 2003; Kennedy *et al.*, 2005). Bipolar disorder appears to be on the rise in adolescents (Narrow *et al.*, 2002), with its onset occurring at ever-earlier ages, and it is associated with substance abuse in over 50 percent of cases (Papolas and Papolas, 2002). Mania and schizophrenia are linked both epidemiologically (Berrettini, 2000; Papolas and Papolas, 2002) and in terms of their symptoms, as they can be extremely difficult to tell apart in their acute phase and can in combination receive the diagnosis "schizoaffective." Bipolar disorder also exhibits a fairly high co-morbidity with attention-deficit/hyperactivity disorder, obsessive-compulsive disorder and substance abuse, with estimates ranging from 15 percent to over 50 percent (Angst *et al.*, 2005; Freeman *et al.*, 2002; Geller *et al.*, 2004; Goodwin and Jamison, 1990; Stahlberg *et al.*, 2004). Bipolar disorder is believed to occur in close to 10 percent of autistic children and is frequently present in family members of autistic children (Papolas and Papolas, 2002; Stahlberg *et al.*, 2004). It is also more frequently found in Tourette's syndrome (5–10 percent) but to a lesser extent than in the other hyperdopaminergic disorders (Kerbeshian *et al.*, 1995).

Bipolar disorder is believed to have a strong genetic component, with heritability estimates from 65 to 85 percent (Bauer and Pfennig, 2005; Berrettini, 2000). Its largest genetic overlap appears to be with schizophrenia, with several putative genetic loci shared in common (Berrettini, 2000). Prenatal and perinatal factors similar to those found in schizophrenia are also implicated in the incidence of bipolar disorder (Buka and Fan, 1999), including both direct ones (e.g. fetal alcohol exposure) as well as indirect ones (e.g. an excess of winter births) (Torrey *et al.*, 1997). The excess of winter births can plausibly be explained by the higher levels of daylight and heat during the summer months – both of which stimulate dopamine – that coincide with the peak of the second trimester, in which prenatal influences on many mental disorders is greatest (Watson *et al.*, 1999). There is a suggestion that increasing societal stress during adolescence may account for some of the putative rise in bipolar disorder in recent decades, but parental observations suggest predisposing traits are present in bipolar individuals even in infancy.

From a neural perspective, mania is best characterized as a neurochemical disorder caused by an elevation primarily of dopamine (responsible for the elevated goal-directed and other activity) and secondarily of norepinephrine (responsible for, in some cases, an elevated mood). Dopamine, in particular, has been implicated by leading theorists such as Depue and Iacono (1989), Goodwin and Jamison (1990), and Swerdlow and Koob (1987). This is largely based on the propensity of L-dopa and various dopamine agonists to produce hypomania in bipolar patients and the therapeutic benefits of dopamine antagonists and lithium, the leading treatment for bipolar disorder that is believed to stabilize dopamine receptor activity (Goodwin and Jamison, 1990; Swerdlow and Koob, 1987). The stimulation of dopamine by light and high temperatures (Arbisi *et al.*, 1994) helps to explain the cyclicity of bipolar symptoms (mania more common in summer) as well as the aforementioned greater prevalence in winter births. There is no specific anatomical locus for the production of mania, although functional imaging studies suggest that activation of the anterior cingulate gyrus and frontal lobes is most likely to accompany it (Adler *et al.*, 2000; Blumberg *et al.*, 2000). Mania is also more likely to occur during activation of the dopamine-rich left hemisphere (particularly in frontal areas), based on lesion studies involving either the left or right frontal lobes (Blumberg *et al.*, 2000; Cummings, 1997; Cutting, 1990; Goodwin and Jamison, 1990; Joseph, 1999; Previc, 2006).

4.2.5   Obsessive-compulsive disorder

Obsessive-compulsive disorder is a relatively common major neuropsychological disorder, with a lifetime prevalence estimated at 2–3 percent (Angst *et al.*, 2005). Obsessions take the form of recurrent and anxiety-producing thoughts while compulsions consist of repetitive and stereotyped behaviors (e.g. hand-washing) and mental acts (e.g. excessive praying) to ward off obsessions and other disturbing events. Obsessive-compulsive traits are positively associated with dopaminergic hyper-religiosity (Previc, 2006), but whether compulsions manifest themselves as religious or not depends to a great extent on the religiosity of the society as a whole. Obsessive-compulsive disorder is moderately to strongly linked to all of the other hyperdopaminergic disorders except Huntington's disease. It has high to very high (>30 percent) co-morbidities with autism, attention-deficit/hyperactivity disorder, bipolar disorder (see earlier sections) and Tourette's syndrome (Como *et al.*, 2005; Faridi and Suchowersky, 2003), and it also has a substantial co-morbidiity with schizophrenia at ~15 percent (Eisen and Rasmussen, 1993; Fabisch *et al.*, 2001).

In contrast to the other hyperdopaminergic disorders reviewed in this chapter, females are slightly more likely to be afflicted by obsessive-compulsive disorder, although symptoms in males tend to develop sooner and are more chronic (Castle *et al.*, 1995; Noshirvani *et al.*, 1991). Males also tend to be generally better represented among those with obsessive-compulsive spectrum disorders, a loosely defined group of disorders that according to some theorists also includes body dysmorphic disorders (dissatisfaction with body appearance), trichotillomania (hair-pulling), and impulse-control disorders (e.g. pathological gambling and sexual addictions) (see Angst *et al.*, 2005; McElroy *et al.*, 1994). However, even in the obsessive-compulsive spectrum disorders, the male bias is not as prominent as in attention-deficit/hyperactivity disorder, autism, schizophrenia, and Tourette's syndrome. As already reviewed, obsessive-compulsive disorder and autism bear an especially strong association with each other, and the relationship between obsessive-compulsive and bipolar disorder is also very strong. Some researchers even argue that impulse-control disorders, which are typically distinguished from classic obsessive-compulsive symptoms by less planning and anti-anxiety intent, are almost always found in mania (Moeller *et al.*, 2001). Despite its rather substantial lifetime prevalence, there is surprisingly little firm evidence concerning the etiology of obsessive-compulsive disorder and whether its prevalence is changing, although it does appear to be more common than previously believed (Szechtman *et al.*, 1999). While there appears to be a heritable component, especially in patients with overlapping Tourette's syndrome, no specific genes have been identified, and there is also little evidence of a prenatal or birth influence on the incidence of this disorder.

There are no gross anatomical pathologies consistently found in obsessive-compulsive disorder, although there does appear to be over-activation in the ventromedial dopaminergic system, including the anterior cingulate and portions of ventromedial (orbital) frontal cortical regions (Adler et al., 2000; Rosenberg and Keshavan, 1998). In fact, removal of the anterior cingulate in a procedure known as a cingulotomy is the leading surgical technique used to control intractable obsessive-compulsive disorder. Prefrontal cortical regions, including the lateral prefrontal cortex, function fairly normally in obsessive-compulsive disorder (Abbruzzese et al., 1995), but they do not appear to provide sufficient inhibition over the subcortical and medial dopaminergic systems, which are primarily responsible for the debilitating perseverative and ritualistic behaviors.

As in schizophrenia, the average age of onset of obsessive-compulsive symptoms is in the early twenties, and obsessive-compulsive symptoms may be precipitated by stress and anxiety, which depletes serotonin and elevates dopamine in the mesolimbic areas (Finlay and Zigmond, 1997). The dopaminergic excess may be especially important in accounting for the broader obsessive-compulsive spectrum, including sexual, gambling, video-game, and other psychological addictions. For example, video-game playing stimulates striatal dopamine release (Koepp et al., 1998), while dopamine treatment for Parkinson's disease increases pathological gambling (Dodd et al., 2005). The dopaminergic excess is also consistent with evidence for a left-hemispheric overactivation in obsessive-compulsive disorder (Otto, 1992), as brain imaging studies have shown that D_2 receptor binding – indicating the amount of receptors not already occupied by dopamine – is significantly reduced in the left hemisphere of obsessive-compulsive patients relative to controls (Denys et al., 2004). As in other hyperdopaminergic disorders, the excessive behavioral activity in obsessive-compulsive disorder is best blocked by a combination of anti-dopaminergic drugs and serotonin re-uptake blockers (Carpenter et al., 1996; Szechtman et al., 1999), which may be most effective in reducing the specific left-hemispheric activation (Benkelfat et al., 1990).

4.2.6   Parkinson's disease

Parkinson's disease, first described by James Parkinson in 1817, is a neurological disorder in which degeneration of nigrostriatal dopaminergic neurons leads to a dramatic reduction in dopamine levels in the striatum (Litvan, 1996). Behaviorally, Parkinson's disease is characterized primarily by a progressive loss of voluntary motor functions along with tremors and secondarily by a set of intellectual deficits initially confined to executive-type ones (Previc, 1999). Parkinson's disease is one of the most common of neurological disorders with a prevalence of ~1–2 per 1,000 and ten times that in those over sixty-five. Parkinson's disease is believed to be mainly caused by the malfunction of Parkin and various other neuroprotective genes along with exposure to environmental toxins (Corti et al., 2005). Although dopamine loss is a cardinal feature of aging human brains in general (Bäckman et al., 2006), the selective dopamine reduction is much more severe in Parkinson's disease.

The selective dopamine loss in Parkinson's disease has led researchers to use it as a model for understanding the role of striatal/lateral-frontal dopamine in motor behavior and cognition The deficits in voluntary motor behavior extend to all types of movements but, as noted earlier, they do not include automatic or reflexive behaviors. The cognitive impairments initially appear restricted mainly to executive intelligence, including deficits in:

1. maintaining and operating on items in working memory (as opposed to short-term memory per se, since digit span is not impaired);
2. shifting conceptual sets, as required by the Wisconsin Card-Sorting Test described in Chapter 3;
3. sequential ordering of responses; and
4. maintaining goal-oriented scanning (see Previc, 1999, for references).

The motor deficits are more severe in upper space, as Parkinson's patients have trouble making upward eye movements and display downward errors in their arm trajectories in the absence of vision (see Corin et al., 1972; Poizner and Kegl, 1993). The upward deficits are, of course, consistent with the role of dopamine in mediating interactions in the most distal portions of 3-D space (Previc, 1998). It should be noted that the symptoms of Parkinson's disease initially appear on the left side of the body (controlled by the right hemisphere) in most patients (Previc, 1991), which is presumably due to the fact that dopamine levels are lower to begin with in the right hemisphere of most individuals.

The principal treatment for Parkinson's disease is the administration of drugs that increase dopamine output, mainly the dopamine precursor L-dopa. But, while L-dopa obviously helps to ameliorate or at least retard the progression of classic Parkinson's deficits involving motor initiation and sequencing and working memory, it also leads to motor problems (dyskinesias), affective disturbances (e.g. depression) and even psychotic (hallucinatory) behavior in many Parkinson's patients (Jankovic, 2002). The dyskinesias produced by L-dopa can resemble the choreas and dystonias found in the hyperdopaminergic state associated with Huntington's disease. Disturbances of sleep and cardiovascular and temperature control (increased sweating) after L-dopa therapy (Quadri *et al.*, 2000) are consistent with the mostly parasympathetic actions of dopamine, as reviewed in Chapter 2.

4.2.7 Phenylketonuria

Phenylketonuria is a disorder involving an absence or mutation of a single gene (PAH) responsible for production of the enzyme phenyl-alanine hydroxylase, which converts phenylalanine to tyrosine (Kenealy, 1996). In the absence of a normal PAH gene, excessive phenylalanine and diminished tyrosine occur and combine to produce severe mental retardation. The major cause of the subnormal intelligence is the excessive phenylalanine, which disrupts basic brain and skeletal development, including the formation of myelin. Studies have shown a decrease of one-half standard deviation in intelligence scores for each 300 µmol/L of phenylalanine in the blood, with levels >1,200 µmol/L producing severe retardation (Burgard, 2000).

Although abnormalities in the PAH gene itself are not rare, afflicting ~ 2 percent of the population, the recessive nature of the gene tends to diminish the actual incidence of the phenotype to about 1 in 10,000 among Caucasians. Because high concentrations of phenylalanine result in detectable increases in its metabolite – phenylpyruvic acid – untreated phenylketonuria is now quite rare in the developed world. The accepted treatment for phenylketonuria is dietary restriction of phenylalanine in early childhood, and tyrosine supplements serve to compensate for the failure to convert phenylalanine to tyrosine. Restricted phenylalanine alone does not prevent all intellectual deficits – especially those associated with executive functions like working memory and cognitive-shifting – but neither does supplemental tyrosine alone prevent the majority of mental retardation. It has also been shown that phenylketonuria during pregnancy (known as "maternal phenylketo-nuria") produces intellectual deficits in offspring not even genetically prone to this disorder (Hanley *et al.*, 1996; Kenealy, 1996).

The detrimental lack of tyrosine during early development is con-sistent with the importance of tyrosine to the synthesis of dopamine and norepinephrine. In particular, the executive deficits in phenylketonuria have been attributed to dysfunction of the lateral prefrontal dopami-nergic systems (Diamond *et al.*, 1997; Welsh *et al.*, 1990), presumably resulting from the decreased tyrosine. This is consistent with the effects of tyrosine restriction during pregnancy on dopamine levels and behavior in animals (Santana *et al.*, 1994) and with the deleterious effect of maternal iodine restriction and hypothyroidism on the conversion of tyrosine to dopa and on offspring intelligence in humans (see Previc, 2002). Because the altered neurochemistry in phenylketonuria is less specific for dopamine than is the nigrostriatal degeneration in Parkinson's disease, many of the motor and other deficits found in the latter disease are not manifested in phenylketonuria, although a tendency toward hyperthermia may be present in both disorders (Blatteis *et al.*, 1974; Quadri *et al.*, 2000).

4.2.8 Schizophrenia

Although it afflicts only about 1 percent of the population, schizophrenia is the best-studied of all neuropsychological disorders, partly because of the bizarre thought patterns and delusions associated with it and the fact that so many great minds have at least temporarily succumbed to it or similar psychoses, including the physicists Newton and Faraday (see Karlsson, 1974), the playwright August Strindberg, the novelist Franz Kafka, the poet Ezra Pound, and the Nobel laureate and mathematician, John Nash. Recent meta-analyses have determined that there is an overall male excess in schizophrenia of ~40 percent (Aleman *et al.*, 2003), but early-onset schizophrenia (the most severe form) is about twice as likely to occur in males as in females. By contrast, the later-onset, milder version is more likely to occur in females in middle age as the inhibitory influence of estrogen over dopamine begins to wane with approaching menopause (Castle, 2000). As previously reviewed, schizophrenia is strongly associated with obsessive-compulsive disorder and mania, and its psychotic symptoms are similar to those found in Huntington's disease. Schizophrenia also bears a smaller but still greater-than-expected co-morbidity with attention-deficit/hyperactivity disorder and autism and probably Tourette's syndrome (Muller *et al.*, 2002), although little formal data exist concerning the last connection.

The major diagnostic signs in schizophrenia are categorized as either "positive symptoms" (e.g. hallucinations, delusions, thought disorder) or "negative symptoms" (affective disturbances such as poor social interaction, depressed mood, and anhedonia, the loss of pleasurable sensations). The "split" conveyed by the name schizophrenia does not refer to a divided self as is often popularly conveyed but rather to an inner world divorced from external reality. Some of the specific thought deficits in schizophrenia include:

1. insensitivity to situational context/feedback, reflected in the lack of effects of previous reward contingencies (e.g. latent inhibition, as

described in Section 3.2.3) and in difficulties in filtering out irrelevant thoughts (Swerdlow and Koob, 1987; Weiner, 2003);
2. remote (loosened) associations between seemingly unrelated stimuli; and
3. a reduced ability to understand the intent of others.

All of these symptoms reflect to varying degrees a hyperdopaminergic state comprised of insufficiently grounded and controlled mental activity (Previc, 2006; Swerdlow and Koob, 1987; Weiner, 2003). While certain features of schizophrenia are also similar to those of autism (deficits in global processing and theory of mind), mania (delusions), and Huntington's disease (psychosis), these and other hyperdopaminergic disorders are also characterized by increases and/or abnormalities in motor output (e.g. stereotypical movements), whereas disorganized thoughts are more salient than aberrant motor behavior in schizophrenia, at least early on in the illness. In the acute phase, schizophrenia cannot easily be distinguished from mania, but the latter is more transient and is not associated with the negative schizophrenic symptoms, which tend to develop more chronically. Without neurological testing, schizophrenic psychosis is also difficult to distinguish from the psychosis found in temporal-lobe epilepsy, in which dopamine is elevated (Previc, 2006). Similarly, milder delusional tendencies and unusual sensory experiences are found in a normal personality variant known as "schizotypy," which is often associated with mild neurological abnormalities of the temporal lobe (Dinn et al., 2002). Psychosis is also found in 10–20 percent of hyperthyroid patients (Benvenga et al., 2003), which is interesting in view of the evolutionary link between thyroid output and dopamine that will be discussed in Chapter 5.

Prominent thought disorders in schizophrenics include paranoia and delusions of control, ranging from the "Messiah complex" to a belief that one is being controlled by external forces such as aliens. Religious themes also figure prominently among the schizophrenic delusions, but these are replaced by nonreligious (e.g. sexual or grandiosity) themes in less religious societies (Previc, 2006). Schizophrenic delusions are often cosmic in nature, as evidenced in the following description by Jaspers (1964: 295):

The cosmic experience is characteristic of schizophrenic experience. The end of the world is here, the twilight of the gods. A mighty revolution is at hand in which the patient plays a major role. He is the center of all that is coming to pass. He has immense tasks to perform, of vast powers. Fabulous distant influences, attractions, and obstructions are at work. "Everything" is always involved: all the peoples of the earth, all men, all the gods, etc. The whole of human history is experienced at once. The patient lives through infinite millennia. The instant is an eternity to him. He sweeps through space with immense speed, to conduct mighty battles; he walks safely by the abyss.

The schizophrenic emphasis on religious and cosmic themes reflects a fundamental disturbance in their interactions with 3-D space and in the systems that deal with 3-D space (see Previc, 1998, 2006). Relative to normals, schizophrenics tend to show a bias toward extrapersonal space and a deficit in peripersonal spatial operations. This extrapersonal bias is reflected in numerous ways, including:

1. a flattened 3-D appearance of the world, which indicates a lack of depth perception ordinarily provided by our peripersonal system;
2. upward deviations of the eyes known as "oculogyric crises";
3. an upper-field predominance of visual hallucinations;
4. deficits in pursuit tracking of objects, which is used mostly in peripersonal space;
5. deficits in body imaging and awareness;
6. loss of prosody, emotional perception, and other functions that rely on our body-arousal system; and
7. reduced sensitivity to bodily signals, such as from the vocal musculature, which leads to the erroneous attribution of self-generated internal thoughts to external voices (Bick and Kinsbourne, 1987).

Even the loosened, remote thought associations characteristic of schizophrenics may be a manifestation of a more fundamental tendency to connect spatiotemporally distant stimuli (Previc, 2006).

The emphasis on extrapersonal themes is consistent with the functional neuroanatomical profile in schizophrenia. Schizophrenia is generally believed to reflect overactivity in the ventral cortical and limbic pathways – particularly the medial-prefrontal and medial-temporal areas and associated subcortical regions comprising the action-extrapersonal system – along with diminution of parietal and occipital inputs (see Previc, 2006). The hallucinations, delusions, and other positive symptoms of schizophrenia emanate primarily from activity in the medial dopaminergic pathways, while the negative symptoms tend to reflect reduced activity in the parietal lobe and other posterior areas (Buchanan et al., 1990). However, even positive symptoms such as hallucinations depend to some extent on diminished posterior sensory inputs. For example, all of us tend to hear or imagine things in our brains, but the discrimination of internally generated auditory and visual signals from external reality would be more difficult if we could not feel ourselves talk or if we did not receive normal auditory and visual signals from the environment.

Although some models of schizophrenia posit that medial-temporal inputs are reduced relative to overactive medial-frontal structures (e.g. Weiner, 2003), overactivation of the temporal lobes is more likely as suggested by the aforementioned resemblance of schizophrenia to temporal-lobe epilepsy, in which the medial temporal lobe is hyper-excitable (Sachdev, 1998). As noted in Chapter 3, lateral prefrontal regions may be less active relative to medial subcortical circuits (Abi-Dargham and Moore, 2003; Bunney and Bunney, 2000; Davidson and Heinrichs, 2003), thereby preventing the latter's ego-control mechanisms from overcoming the chaotic thoughts inspired by medial dopaminergic activity. Overactivation of the left temporal lobe in schizophrenia has also been well-documented (Cutting, 1990; Gruzelier, 1999; Gur and Chin, 1999; Rotenberg, 1994), consistent with the general role of the dopamine-rich left hemisphere in delusions and hallucinations and with its exaggerated emphasis on extrapersonal space (Previc, 1998, 2006). Few if any of these functional imbalances in brain activity directly result from neuroanatomical pathology; indeed, aside from a possible reduction of volume in the temporal lobe, there is little evidence that actual neuroanatomical damage contributes to or even correlates with schizophrenia (Chua and McKenna, 1995). Hence, the consensus of researchers is that the schizophrenic behavioral syndrome is mostly a consequence of functional changes in specific neurochemical circuits, principally those involving dopamine.

The theory of dopamine overactivation in schizophrenia is one of the longest standing and most widely accepted in neuropsychology (see Kapur, 2003; Swerdlow and Koob, 1987). Dopaminergic elevation can account for every positive symptom of schizophrenia, ranging from the loosened thought associations to the saccadic intrusions to the delusions and hallucinations and even the bias toward distant space. Drugs that elevate dopamine like L-dopa and amphetamine create or worsen psychotic symptoms, whereas drugs that block dopamine activity (mostly D_2 receptors) have long been the treatment of choice in this disorder. Other neurochemical systems, especially serotonin and glutamate, are also implicated in schizophrenia (Vollenweider and Geyer, 2001). Both of these systems inhibit dopamine and, when blocked, are believed to contribute to hallucinations and other positive symptoms (see Previc, 2006). Their involvement in schizophrenia is attested to by the greater therapeutic effectiveness of newer "atypical" antipsychotic drugs such as clozapine, which not only block dopamine D_4 receptors but also affect a variety of other neurochemical systems. As noted in earlier chapters, serotonin appears to be mostly involved in bodily arousal and peripersonal functions, so its reduction would also be expected to tip the balance toward extrapersonal activity (Previc, 1998, 2006).

Like most other neurodevelopmental disorders involving dopamine, schizophrenia is caused by a combination of genetic, prenatal, perinatal, and postnatal influences (Lewis and Levitt, 2002). The concordance among identical twins is about 50 percent, which suggests a considerable genetic influence, but no single gene or genetic factor has consistently been shown to be linked to schizophrenia (Lewis and Levitt, 2002). Along with hypoxia at birth, maternal infection/fever and malnutrition – especially in the second trimester of pregnancy – represent two of the best-documented prenatal factors (Watson *et al.*, 1999), with maternal infection leading to as much as a seven-fold increase in the risk of schizophrenia (Brown, 2000). All of these maternal factors elevate brain dopamine, and the delayed prenatal influence (second trimester and beyond, consistent with a well-established excess of winter births; Torrey *et al.*, 1997) suggests that schizophrenia, even more than autism, may involve dopaminergic abnormalities in the later-developing cerebral cortex. Postnatal psychosocial and other stressors that deplete the brain of norepinephrine and serotonin and thereby shift the neurochemical balance further in favor of dopamine are believed to help precipitate most cases of schizophrenia. For example, thermal stress, which leads to elevated dopamine levels (see Previc, 1999), can exacerbate schizophrenia (Hare and Walter, 1978) as well as mania and Tourette's syndrome (Lombroso *et al.*, 1991; Myers and Davies, 1978). However, although various postnatal stressors may serve as catalysts, the fundamental predisposition to schizophrenia clearly arises from genetic as well as early neurodevelopmental influences during the prenatal and perinatal periods (Lewis and Levitt, 2002).

4.2.9   Tourette's syndrome

Tourette's syndrome is a chronic neurodevelopmental disorder characterized by motor tics involving mainly the orofacial region. These tics consist of shoulder shrugging, grimacing, blinking, grunting and even more complex vocal behavior such as echolalia (repeating other's words) or coprolalia (socially inappropriate verbal utterances) (Faridi and Suchowersky, 2003). Tourette's syndrome is associated with poor academic achievement and social adjustment, but it is not nearly as debilitating as other hyperdopaminergic disorders such as autism and schizophrenia. Tourette's syndrome afflicts up to 2 percent of the population (Faridi and Suchowersky, 2003; Robertson, 2003), which is much more common than once believed, and it has a moderate male bias (at least 1.5:1). As reviewed earlier, Tourette's syndrome has extremely high co-morbidities with attention-deficit/hyperactivity disorder (estimates range from 8 percent to 80 percent), autism (up to 80 percent of autistic persons have tics, according to Gillberg and Billstedt, 2000), and obsessive-compulsive disorder (50 percent of Tourette's patients have obsessive-compulsive symptoms, according to Como et al., 2005). Tourette's syndrome exhibits a smaller but still greater-than-expected association with mania and schizophrenia (<10 percent). Finally, Tourette's syndrome is closely related to stuttering, another hyperdopaminergic disorder characterized by vocal dystonias and a male predominance (Abwender et al., 1998). For example, stuttering is accompanied by motor tics in about 50 percent of cases, and it has a similarly high co-morbidity with obsessive-compulsive disorder (Abwender et al., 1998).

Tourette's syndrome has strong familial links and is generally viewed as having a substantial genetic component, with a concordance rate of >50 percent for identical twins (Faridi and Suchowersky, 2003). Genes that regulate dopamine (such as dopamine beta-hydroxylase) have been implicated (Comings et al., 1996), although no particular gene has been identified as critical and so a polygenic influence is suspected (Faridi and Suchowersky, 2003; Nomura and Segawa, 2003). Prenatal factors may also be involved, as suggested by greater maternal than paternal transmission (Faridi and Suchowersky, 2003), but they do not seem to be as influential as in autism and schizophrenia. While tic expression in Tourette's patients can be exacerbated by both physical and psychosocial stress, postnatal/environmental factors also do not seem to play as important an etiological role as in mania and schizophrenia.

Tourette's syndrome is generally viewed as a disorder of the dopamine-rich basal ganglia and the prefrontal, orbitofrontal, and limbic cortical areas. There are no striking neuroanatomical abnormalities as in Huntington's disease, although volume changes in the basal ganglia have occasionally been reported. Although some dopamine-rich brain areas, particularly those in the basal ganglia, may be underactive in Tourette's patients, these areas may paradoxically be overactivated when actual symptoms such as tics are exhibited (see Nomura and Segawa, 2003). The transient dopaminergic overactivation may arise from a supersensitivity of dopaminergic receptors, possibly caused by chronically low striatal dopamine levels (Nomura and Segawa, 2003). The leading treatment for Tourette's syndrome consists of dopamine-blocking drugs such as the typical or atypical neuroleptics (Faridi and Suchowersky, 2003), although other transmitter systems that indirectly affect dopamine levels have also been the subject of treatments. Dopamine agonists at low doses may be of some benefit (Nomura and Segawa, 2003), although amphetamine and similar dopamine-activating drugs tend to increase tic frequency and severity at higher dosages.

4.3   Summary

The preceding review of nine dopamine-related disorders illustrates a very different set of symptoms in disorders that elevate dopamine versus those that reduce it. In the ones in which dopamine is overactive, increased motor activity and/or mental activity is present, whereas slower motor and/or mental activity are found in phenylketonuria and Parkinson's disease. The hyperdopaminergic disorders tend to show strikingly high co-morbidities in most cases, except for Huntington's disease, which is caused by a single-gene mutation. These disorders tend not to be causally related to any overriding neuroanatomical pathology, again with the exception of Huntington's disease. Rather, they

1. mostly reflect serotonergic underactivation versus dopaminergic overactivation in the ventral cortical areas and in the already dopamine-rich left hemisphere;
2. are triggered by stress or anxiety, which is known to increase activity in the ventromedial dopaminergic pathways (see Chapter 2); and
3. are preferentially treated by a regime that involves serotonin boosters or dopamine blockers.

The hyperdopaminergic disorders mostly exhibit a mild to strong male prevalence, at least in early-onset cases, and their overall prevalence is either definitely or possibly rising except in schizophrenia. Consequently, it may be appropriate to think of the hyperdopaminergic disorders not as separate syndromes but rather as *overlapping symptom sets with the same underlying neurochemical imbalances*.

As Table 4.1 indicates, the hyperdopaminergic disorders are hardly monolithic in either their symptoms or etiology. Whereas Huntington's disease is a single-gene disorder and five others (autism, attention-deficit/hyperactivity disorder, bipolar disorder, schizophrenia, and Tourette's syndrome) have varying degrees of genetic inheritance associated with them, four of those with genetic etiologies also show substantial prenatal/perinatal inheritance. And, at least two of the later-onset disorders – schizophrenia and obsessive-compulsive disorder – may be

Table 4.1 *Features of the major hyperdopaminergic disorders.*

Disorder	DA (lateral)	DA (medial)	5-HT	LH overactivation	Male bias	Hyper kinetic	Prenatal/ perinatal	Prevalence
ADHD	?	↑	?	+	3:1	+	*	↑
Autism	↑	?	↓	+	4:1	+	**	↑
OCD	?	↑	↓	+	earlier onset	+	?	?
Mania (bipolar)	↑	↑	?	+	earlier onset, <2:1	+	*	↑
Schizophrenia	?	↑	↓	+	1.4:1	?	**	x
Tourette's	↑	?	?	?	1.5:1	+	?	?

Notes:

⁺for LH indicates predominant symptoms are those of left hemisphere.

⁺for hyperkinetic indicates this symptom is present.

*indicates mild prenatal influence.

**indicates strong prenatal influence.

↑indicates elevation or increase.

↓indicates reduction.

x indicates no change.

? indicates association meriting further research.

Mild male bias refers to ratio of males to females.

Table 4.2 *Co-morbidity of the major hyperdopaminergic disorders.*

Disorder	ADHD	Autism	OCD	Mania (bipolar)	Schizophrenia	Tourette's
ADHD	–	**	***	***	*	***
Autism		–	**	*	*	**
OCD			–	***	**	***
Mania (bipolar)				–	**	*
Schizophrenia					–	*
Tourette's						–

Notes
*represents mild or familial linkages only;
**represents substantial co-morbidity of 10–30 percent;
***represents strong co-morbidity of over 30 percent.

precipitated by underlying trait anxiety or stress that depletes the brain of serotonin and norepinephrine. Some of the disorders (specifically, Huntington's disease and Tourette's syndrome) result from subcortical dopaminergic activation, one (autism) bears some similarities to the over-focusing associated with prefrontal dopaminergic activation, and two others (schizophrenia and attention-deficit/hyperactivity disorder) arguably result from an excess of ventromedial dopaminergic activity relative to lateral prefrontal dopaminergic activity. In two other disorders – obsessive-compulsive disorder and mania/hypomania – over-activation of the ventromedial pathways may co-exist with normal or even elevated lateral prefrontal activity, resulting in heightened behavioral output involving planning and strategy. Another way of viewing the hyperdopaminergic disorders is that they lie on a continuum ranging from the impulsive (e.g. attention-deficit/hyperactivity disorder, mania, some addictions, and schizophrenia) to the compulsive (e.g. autism, obsessive-compulsive disorder). In the former disorders, the medial dopaminergic drive appears greater than the lateral dopaminergic inhibition, resulting in impulsive, disordered but also creative thoughts, while in the latter disorders the lateral and other prefrontal dopaminergic inhibition is relatively intact or even elevated, resulting in over-focusing of behavior and stereotypy.

It is interesting to note that of the various hyperdopaminergic disorders, obsessive-compulsive disorder is the most interrelated, exhibiting substantial or strong co-morbidity with every other hyperdopaminergic disorder except Huntington's disease (Table 4.2). Its high co-morbidities are not surprising in that obsessive-compulsive symptoms most clearly epitomize dopaminergic overactivation – heightened motor or mental activity, goal-directed behavior, concern about spatial and temporally distant events (e.g. the future), and attempts to control the environment in elaborate and extreme ways to ward off emotional or other types of aversive consequences, sometimes to the point of bodily neglect. Furthermore, both the lateral and ventromedial dopaminergic systems are either active or overactive in obsessive-compulsive disorder, and an excess of left-hemispheric activity occurs in it as in almost all of the other hyperdopaminergic disorders. And, while obsessive-compulsive disorder itself shows only a minor gender difference, many obsessive-compulsive spectrum disorders such as gambling, sexual addiction, and excessive video-game use are much greater in males, in line with the general male predominance in the hyperdopaminergic disorders.

Perhaps the most troubling aspect about the hyperdopaminergic disorders is that, except for Huntington's disease and schizophrenia, they are either definitely or possibly rising, which cannot be reconciled with the notion that they are mostly genetically determined. Because prenatal factors clearly influence the risk for developing most hyperdopaminergic disorders, the increasing prevalence of these disorders in modern industrialized nations may more properly be ascribed to demographic changes, deleterious environmental exposures, and/or psychological pressures in modern society that have pushed maternal dopamine levels to dangerously high levels (Previc, 2007), which I will further expound upon in Chapter 7.

5   Evolution of the dopaminergic mind

It is customary, albeit limiting, to view human brain evolution in terms of the events leading up to genetically and anatomically modern humans, now believed to be approximately 200,000 years ago. All subsequent changes in humans are attributed to the effects of "culture," and human "history" is relegated to even more modern events beginning with the formation of agricultural societies. If, however, the expansion of dopamine was not due mainly to genetically mediated changes in our neuroanatomy but rather to epigenetic changes in our neurochemistry, then the physical brain evolution of modern humans has continued all the way to the present. This chapter will focus on the evolution of the dopaminergic mind leading up to, and including, the cultural explosion in *Homo sapiens* that has been termed the "Big Bang" (Mithen, 1996), which occurred first in Southern Africa between 70,000 and 80,000 years ago and later appeared in Europe around 40,000–50,000 years ago, while Chapter 6 will focus on the changes in the dopaminergic mind since the dawn of history. I will highlight two major events in human evolution – the evolution of the "protodopaminergic" mind beginning around two million years ago and the emergence of the later dopaminergic mind with its distinctly human intellectual abilities less than 100,000 years ago. First, however, it is necessary to describe further the contribution of epigenetic influences to inheritance, given the crucial role they appear to have played in our intellectual evolution.

5.1   The importance of epigenetic inheritance

As discussed in Chapter 1, there are strong reasons to believe that the evolution of human intelligence did not depend on changes in brain size or on changes in the genome, at least in the later stages, but rather on an expansion of dopaminergic systems in the brain. What, then, caused the dopaminergic expansion that led to the modern human mind? As alluded to in Chapter 1 and detailed in an earlier theory (Previc, 1999), the evolution of the dopaminergic mind depended on physiological influences and dietary changes that together led to increased dopamine in the brain. Whereas the genome was once believed to almost exclusively determine our inheritance, it is now widely accepted that epigenetic influences, especially those occurring in the womb, affect and sometimes even override gene expression at all levels and thereby modify brain development (Gottlieb, 1998; Harper, 2005; Keller, 2000; Lickliter and Honeycutt, 2003; Nathanielsz, 1999; Petronis, 2001). The maternal environment affects immune function, heart disease, diabetes, and cancer risk (Nathanielsz, 1999; Petronis, 2001) and even gross physical appearance in some cases (Gottlieb, 1998); but most relevant to this thesis, it has been shown to have an especially powerful influence on brain development and behavior, and may even be considered a source of speciation (Lickliter and Honeycutt, 2003). What is critical from the standpoint of human brain evolution is that maternal effects are *trans-generational* in that, for example, a mother with high dopamine levels can prenatally pass those levels on to her children and, in turn, they to their children (Harper, 2005; Lickliter and Honeycutt, 2003; Nathanielsz, 1999).

Three of the most important examples of epigenetic influences on brain development are the effects of prenatal iodine-deficiency (Boyages and Halpern, 1993; DeLange, 2000), the effects of poorly controlled phenylalanine and tyrosine levels in phenylketonuria (Hanley *et al.*, 1996), and the effects of high maternal dopamine levels on the increased risk for autism (Previc, 2007). Prenatal iodine-deficiency syndrome is more likely to affect brain development than skeletal growth and is estimated to place over one billion humans at risk worldwide for reduced intelligence (DeLange, 2000). In autism and phenylketonuria, the maternal risk involves both genetic and nongenetic factors, but even the genetic risk is partly manifested in an aberrant prenatal neurochemical environment. For example, double allelic deletion of the gene responsible for creating the enzyme dopamine beta-hydroxylase that converts the neurotransmitter dopamine to norepinephrine should theoretically produce an individual with no norepinephrine and high levels of dopamine, which not only would disturb mental health (see Chapter 4) but would also be catastrophic physiologically since norepinephrine is the key neurotransmitter maintaining normal sympathetic cardiac output. Even in this case, however, normal maternal norepinephrine levels can overcome the offspring's own lack of norepinephrine (Thomas *et al.*, 1995), whereas conversely an abnormally high maternal dopamine-to-norepinephrine ratio can create excessive dopamine levels even in offspring whose own dopamine beta-hydroxylase gene is normal (Robinson *et al.*, 2001).

Dopaminergic systems in humans are highly susceptible to a myriad of prenatal influences, based on direct and indirect manipulations of maternal dopamine levels during pregnancy. Many studies have shown that maternal ingestion of tyrosine (the precursor to dopa and dopamine), cocaine (which blocks the re-uptake of dopamine), amphetamine (which both stimulates dopamine release and blocks dopamine re-uptake), and haloperidol (which blocks the action of dopamine postnatally) all affect postnatal dopamine levels and various dopamine-mediated behaviors in offspring (e.g. Archer and Fredriksson, 1992; Santana *et al.*, 1994; Zhang *et al.*, 1996). The general finding from these and other studies is that stimulation of maternal dopamine systems prenatally results in enhanced postnatal dopaminergic activity (Santana *et al.*, 1994), whereas reduced maternal dopamine activity diminishes postnatal dopaminergic activity (Zhang *et al.*, 1996).

As noted in Chapter 4, the importance of the prenatal environment to brain development has ever challenged the basic assumptions of behavioral genetics. It is assumed that monozygotic and dizygotic twins share the same prenatal environment and that the difference between their concordance rates is genetic; whereas, a difference in concordance rates for dizygotic versus regular siblings is caused by the greater shared prenatal and/or postnatal environments of the former. Using these methods, the genetic influence has been estimated at ~50 percent in the case of intelligence (Dickens and Flynn, 2001) and even higher in some of the major psychological disorders like attention-deficit/ hyperactivity disorder, autism, bipolar disorder, and schizophrenia. In reality, though, the prenatal and postnatal environments of mono-zygotic and dizygotic twins are not identical and the importance of shared zygosity is probably greatly exaggerated (Mandler, 2001; Prescott *et al.*, 1999).

It is not clear exactly how much of the dopaminergic expansion during hominid evolution can be ascribed to epigenetic factors versus how much can be ascribed to genetic adaptations. But, the recent large increases in many industrialized nations in intelligence (Dickens and Flynn, 2001) and in autism (Previc, 2007) – two phenomena clearly tied to dopamine, as discussed in Chapters 3 and 4 – suggest that popula-tion-wide changes in dopaminergic activity can occur and have occurred without changes in the genome. In the remainder of this chapter, I will propose a set of scenarios that provide a plausible and comprehensive explanation for the rise of the dopaminergic mind. These scenarios, which involve both genetic and epigenetic inheritance, will be divided into two main portions – evolution of the protodopaminergic mind from a few million years ago to approximately 200,000 years ago, and evolution of the later dopaminergic mind leading to the cultural "Big Bang" ~70,000–80,000 years ago.

5.2   Evolution of the protodopaminergic mind

5.2.1   *Environmental adaptations in the "cradle of humanity"*

Somewhere between five and six million years ago, the first hominids – the Australopithecines – began to appear in sub-Saharan Africa (Arsuaga, 2003; Coppens, 1996). This date has been established on the basis of both fossil evidence and mitochondrial DNA evidence.[1] There were a series of Australopithecine lineages, only one of which (*Australopithecus afarensis*) is generally accepted as having directly led to humans.

The divergence of humans and chimpanzees occurred within one to two million years of a reactivation of rifting that eventually resulted in the East African plateau becoming considerably drier than the West African tropical forest (Arsuaga, 2003; Coppens, 1996). Fossil evidence now conclusively demonstrates that the early hominids were confined to an open, arid savanna environment that may have surrounded streams, lakes, or other sources of water (Brunet *et al.*, 1995). *Australopithecus* was clearly bipedal, albeit with retained arboreal capabilities (Arsuaga, 2003; Coppens, 1996). Bipedalism would allow the early hominids to exploit the open savanna environment to a much greater extent than the great apes because of their greater ease of locomotion (Carrier, 1984) and because a bipedal posture is much less likely to absorb the heat of the sun (Wheeler, 1985). In turn, the greater exploitation of the savanna niche led to less of a reliance on a frugivorous diet than in the case of the forest-dwelling chimpanzee (Grine and Kay, 1988).[2] Other adaptations to a thermal environment, to be discussed shortly, additionally contributed to the advantage of the early hominids over the great apes in the increasingly arid savannas of East Africa, an advantage that is borne out by the absence of any co-mingled ape fossil remains in the locales of the early hominids (Brunet *et al.*, 1995; Coppens, 1996).

Despite their bipedalism, the australopithecines were short-statured creatures with brains no larger than that of the chimpanzee. A continuing divergence of gracile australopithecines *(Australopithecus*

---

[1] Mitochondrial DNA in cells is, unlike normal DNA, contained in the cell's nucleus, transmitted only through the mother. Based on a known rate of mutation, systematic deviations in DNA among individuals and species can provide reasonably good estimates concerning the time-course of biological evolution (see Cann *et al.*, 1987).

[2] It is worth noting, as evidence of their limited locomotory prowess, that modern apes traverse a total of only about one-half mile per day (Bortz, 1985).

afarenis) from robust australopithecines is believed to have eventually led to the emergence of the genus *Homo* around 2–2.5 million years ago in East Africa (Arsuaga, 2003; Coppens, 1996; Falk, 1990). *Homo habilis* is believed to have represented the first major advance over chimpanzees in relative brain size, with its brain estimated to be about 50 percent larger than that of *Australopithecus* in allometric (i.e. brain-to-body) terms (Arsuaga, 2003; Falk, 1990). *Homo habilis* also represented a major advance culturally, in that a large and diverse array of stone-flaked tools have been found near its remains (Coppens, 1996). These finds have often been located at a distance from where skeletal remains were found, suggesting a transport to animal carcasses from which meat and marrow were removed (Arsuaga, 2003). Such behavior would require a certain amount of foresight and, in turn, an expanded emphasis on more distant space and time. The evolutionary scenario leading from *Australopithecus* to the emergence of *Homo* is not entirely clear, although Falk (1990) for one argues that the gracile australopithecines may have occupied a more open savanna niche than did the robust australopithecines.

What is clear is that the appearance of *Homo habilis* occurred in an era of additional climate change during the transition from the late Pliocene to early Pleistocene geologic epochs, in which the aridity of the sub-Saharan East African plateau further increased (Arsuaga, 2003; Coppens, 1996). A worldwide cooling took place between two and three million years ago, resulting in a lowering of humidity and a dramatic decrease in vegetation; for example, the ratio of tree to grass pollens decreased from 0.4 to 0.01 in the Omo valley of Ethiopia during this period (Coppens, 1996). The increasing dryness of the East African plains favored the inclusion of meat in the diet, which is believed to have occurred around two million years ago (Eaton, 1992; Leonard and Robertson, 1997) as attested to by the animal bones and cutting tools typically found with *Homo habilis* remains and the more efficient chewing capability of its teeth and jaws (Coppens, 1996). It should be noted that present-day hunter-gatherers now consume about one-third of their calories from animal sources (Bortz, 1985; Eaton, 1992; Lee, 1979), which is comparable to that of hunter-gatherers in the Early Stone Age (Eaton, 1992), but more importantly, meat consumption can rise to 100 percent of the diet in times of drought (Bortz, 1985; Lee, 1979).

The emergence of *Homo habilis* was followed by several waves of human migration out of Africa and onto the Eurasian land mass, the first of which may have occurred as early as 1.9 million years ago (Wanpo *et al.*, 1995). By the early Pleistocene era, a relatively advanced *Homo erectus* and its exclusively African cousin (*Homo ergaster*) with much larger brain sizes (~1,000 cc) and height than *Homo habilis* had populated wide areas of Africa, Europe, and Asia. Eventually, *Homo erectus* evolved into archaic humans around 500,000 years ago in Africa, including *Homo helmei* and *Homo heidelbergensis* (found in Africa) and *Homo sapiens neanderthalensis* (also known as the Neanderthals, who later dominated Europe for ~200,000 years and lived contemporaneously with modern humans). The continuing African evolution of modern humans is further supported by numerous fossil and archaeological findings, including:

1. the earliest mixture of *Homo erectus* and *Homo sapiens* cranial features, found in East African skulls dating back approximately one million years;
2. the earliest fossil remains of archaic *Homo sapiens* in sub-Saharan Africa, slightly less than 500,000 years ago;
3. the origin and continued evolution, primarily in Africa during the Early Stone Age (Early Paleolithic Era), of primitive stone tools such as bifacial stone flakes and hand-axes; and
4. the first appearance of blade technology in southern Africa at the beginning of the Middle Stone Age (~250,000–300,000 years ago).

The beginning of the Middle Stone Age appears to have occurred slightly after the split between Neanderthals and modern humans, which is believed to have occurred around 400,000 years ago, based in part on the slight difference (0.5 percent) between the Neanderthal and modern human genomes (Noonan *et al.*, 2006).

It is generally accepted by most anthropologists that modern humans are mostly if not exclusively the product of Africa, which is often referred to as the "cradle of humanity." Recent archaeological and DNA evidence independently confirm that anatomically and genetically modern humans first emerged in Africa sometime around 200,000 years ago. Based on the rate of random mutations in mitochondrial DNA and y-chromosomal DNA and the greater diversity in modern-day DNA in sub-Saharan Africa, divergence analyses suggest that all *Homo sapiens sapiens* evolved from a small gene pool of less than 10,000 in Africa approximately 200,000 years ago (Cann *et al.*, 1987; Hammer, 1995; Templeton, 2002; von Haesler *et al.*, 1996).[3] Although until recently the oldest fully anatomically modern human skull from Klasies River in South Africa dated to 130,000 years ago, a more recent skull from

---

[3] Although specific polymorphisms and other DNA markers point to older dates for the occurrence of the most recent common ancestor of all humans, they are more difficult to interpret in that they are based on much less data and are much more variable, whereas the mitochondrial DNA and y-chromosomal data collectively derive from over 1,500 individuals each and are remarkably consistent in their dating (Templeton, 2002).

Omo Kibish in Ethiopia dating back to 195,000 years ago shows essentially modern features (McDougall *et al.*, 2005), and it is believed that around 150,000 years ago descendants of the earliest *Homo sapiens sapiens* migrated from this region to South Africa (Behar *et al.*, 2008).

It may be concluded, therefore, that conditions in Africa spearheaded virtually every major evolutionary advance in the path from chimpanzees to modern humans. But, many issues pertaining to the evolution of humans have not been conclusively settled, such as how gradual or punctuated that evolution was or how closely our anatomical evolution matched our cognitive evolution. There is much more evidence for gradualism in our anatomical evolution, as continual progression toward the modern human skull occurred from two million to two hundred thousand years ago, in contrast to the rather meager advances in cultural output before 100,000 years ago (Coppens, 1996). Around 200,000 years ago, a striking divergence between our anatomical and cultural evolution occurred in that the modern human genome and craniofacial structure were essentially finalized, whereas over 100,000 years more were required to produce the first clear evidence of art, beads, advanced tools, commerce, and other indicants of a modern-like intellectual capability. Indeed, the cultural distinctions between Neanderthals and genetically modern humans were not all that dramatic in many places such as the Middle East even as late as 90,000 years ago (Arsuaga, 2003; Mellars, 2006; Shea, 2003; Wynn and Coolidge, 2004), and later cultural advances may have occurred in Neanderthals either independently (Zilhão *et al.*, 2006) or after interacting with modern humans in Europe (Amos, 2003; Arsuaga, 2003), despite the former's clearly different genetic makeup and craniofacial anatomy (e.g. larger brain and brow). There is even evidence that modern humans may have been temporarily displaced by Neanderthals in the Levant region (present-day Israel) around this time (Arsuaga, 2003; Mellars, 2006; Shea, 2003).

Whereas modern humans 100,000 years ago did not demonstrate a huge, if any, intellectual advantage over their Neanderthal cousins, this was not true 30,000–40,000 years later. Beginning around 65,000 years ago, there was a rapid littoral expansion of modern humans across the Red Sea into Southwest, South, and Southeast Asia – about four kilometers per year, according to Macaulay *et al.* (2005) – followed by a fairly rapid replacement of the Eurasian Neanderthal populations. A major biologically driven behavioral change presumably occurred in this period to decisively set the stage for the modern human intellect (Shea, 2003). Heretofore, theorists have attempted to understand the origins of the modern human mind in terms of various differences between us and our anatomically and genetically distinct Neanderthal cousins, but a much greater clue to the origins of the human mind lies in a comparison of modern human behavior of 70,000 years ago with that of the genetically and anatomically modern humans of 130,000 years earlier in Africa.

The next section will attempt to explain why sub-Saharan Africa was so pivotal in the evolution of humans during the early-to-middle Pleistocene geologic era spanning the period from two million to 200,000 years ago, particularly regarding the environmental and dietary influences that evidently led to the first major rise of dopamine in hominid evolution.

5.2.2   Thermoregulation and its consequences

The drying of the savannas over the course of millions of years forced a retreat of chimpanzees and gorillas to the lush forests on the western side of the Eastern rift valley. For a small and relatively defenseless creature such as *Homo habilis* that could tolerate heat, a drier climate offered a huge opportunity in that it could exploit the midday environment in which the great apes and many dangerous predators would be relatively inactive, while at night it could retreat to caves and even arboreal safety.

Despite our ability to thrive in a wide variety of tropical environments, humans are still at risk in hyperthermic environments, as illustrated by the large number of heat stroke deaths that occur in the elderly during urban heatwaves and in the young during extreme physical exertion in which sweating may be prevented because of heavy clothing (Figa-Talamanca and Gualandi, 1989). Humans also face a special thermal challenge that nonhuman primates do not, in that very large brains generate much more heat than do smaller ones. However, humans have evolved several highly efficient heat-loss mechanisms that arguably provide us with a greater thermal tolerance than any other species in the animal kingdom (Bortz, 1985; Carrier, 1984). The combination of such traits as a bipedal posture, hairless skin, and an extraordinary number of sweat glands was arguably of much greater survival value to the early hominids than were any of the intellectual advances that occurred in them.[4]

The adoption of a bipedal posture, while primarily beneficial in terms of locomotion and freedom of the hands, also reduces the radiant head load of humans when the sun is at its peak (Wheeler, 1985). The optimization of bipedalism, especially for endurance running, required various improvements in the ability of lower-limb tendons, joints, and

---

[4] Actually, the skin of humans is far from hairless, but the individual hairs are much smaller and finer than in other primates.

bones in humans to be efficient and and durable (Bramble and Lieberman, 2004), while elongating the body produced a concomitant increase in the body's surface-to-mass ratio, which facilitates evaporative heat loss by having relatively more of the human body exposed to air. A large surface-to-mass ratio is a characteristic of all tropical human populations, as predicted by the "Allen's rule" of mammalian thermal physiology (Jablonski, 2004). The hairless skin of humans – unique among primate species – further increases the efficiency of heat dissipation through the skin (Wheeler, 1985; Zihlman and Cohn, 1988). The most important element of our extraordinarily efficient heat loss capability is the enormous number (~2–4 million) and efficiency of eccrine sweat glands found on our bodies. Unlike furry animals that have a large number of apocrine or oily sweat glands (40 percent, in the case of apes), eccrine or watery sweat glands vastly predominate in humans (Jablonski, 2004). Because of these glands, our physical exertion in a desert environment may result in the loss of twelve liters or more of sweat per day (Bortz, 1985), with each liter of sweat carrying away ~600 kcal of heat (Nunneley, 1996). To maintain normal hydration, as much as 15–20 liters of water must be consumed daily while exercising in a hot, dry environment (Nunneley, 1996), which is consistent with the restriction of early hominid fossil remains to the shores of what in the early Pleistocene era were thriving rivers and lakes. For the most part, the heat loss achieved through sweating is under central control rather than being determined by local skin temperatures. This factor is critical because, under exertion in heat, the gradient between environmental temperature and core temperature may limit the heat loss from passive mechanisms like convection and conduction (Jablonski, 2004). It should be noted that eccrine sweating is an efficient means of heat loss only when fur is not present, which strongly suggests the co-evolution of hairlessness, an increase in eccrine sweat glands, and probably bipedalism as well (Folk and Semken, 1991; Jablonski, 2004; Zihlman and Cohn, 1988).

The functional utility of the exceptional adaptation of hominids to heat stress lies in the ability to engage in persistence (chase) hunting and midday scavenging (Bortz, 1985; Bramble and Lieberman, 2004; Carrier, 1984; Krantz, 1968; Shipman, 1986; Zihlman and Cohn, 1988). Chase hunting involves pursuing an animal in the hot sun until it enters a hyperthermic condition and dies ("chase-myopathy"); for instance, zebras and cheetahs will grow hyperthermic to the point of collapse after an all-out chase of about one kilometer (Bortz, 1985; Taylor and Rowntree, 1973). Persistence hunting depends not on a sophisticated hunting technology (which the early hominids almost certainly did not have) but only on physical endurance, and it is still occasionally pursued by modern Bushmen in Africa (Bortz, 1985; Carrier, 1984). The wildebeest of Eastern and Southern Africa is especially prone to chase-myopathy because it is reluctant to leave its area (Taylor, 1980, cited in Bortz, 1985), which is noteworthy because wildebeests are believed to have comprised the single-largest source of meat in the diets of the East African hominids (Bortz, 1985). The only tools needed for chase hunting are primitive stone-cutting instruments used to extricate the meat from the dead animal. Such tools could also have been used in scavenging meat/marrow from carcasses (Shipman, 1986), a thermally demanding task given that the traveling long distances to and from the carcass and the extrication of the marrow presumably occurred during midday hours, when nocturnal predators would have been less of a threat. Our hominid ancestors evidently engaged in both chase-hunting and scavenging, though probably more of the latter (Shipman, 1986).[5]

While the human capacity to dissipate heat is impressive relative to other animals, the use of this capacity to achieve the physical endurances necessary for successful scavenging and chase-hunting requires that several conditions be met. First, the ambient environment must be arid for optimal sweating to occur (Nunneley, 1996). Second, a fluid supply must be readily available, given the loss of hydration caused by the sweating during the endurance activity (Carrier, 1984). Finally, the thermoregulatory system must be able to kick in rapidly during extreme physical exertion, which creates at least twenty times the heat load that is accrued while we are at rest (Nunneley, 1996). The first two of these conditions were fulfilled by the early hominid environments of Eastern and Southern Africa, which were essentially arid savannas that are believed to have contained either nearby lakes or rivers.[6] The last condition could have been met by an expansion of dopaminergic systems in the brain, particularly the dopaminergic nigrostriatal pathway. As noted

---

[5] Regardless of whether the animal was killed or scavenged, the extraction of meat was made easier by an additional increase in the length and flexibility of the thumb (Carroll, 2003), which probably took place as part of the general physiological/anatomical adaptation during the late Pliocene/early Pleistocene transition.

[6] The fact that our early hominid ancestors as well as prehistoric modern humans in sub-Saharan Africa tended to congregate near sources of water does not imply that our ancestors once spent large amounts of time in an actual aquatic environment, as proposed by Morgan (1997) and others. Much of the "aquatic" theory has been discredited (see Moore, www.aquaticape.org), and all of the supposedly aquatic physiological adaptations either do not exist (e.g. humans do not a "diving reflex" like other marine mammals) or were adaptive for other purposes (e.g. a hairless skin helped to dissipate heat and a descended larynx promoted vocal communication).

in Chapter 2, the nigrostriatal system mediates voluntary motor activity, so it is ideally suited to lowering body temperature after the onset of activity (Cox, 1979; Lee *et al.*, 1985). The nigrostriatal stimulation of dopaminergic heat-loss mechanisms evidently begins within minutes of activation of the striatum during motor activity and continues until core temperatures reach about 2–3°C above normal, after which dopamine levels and temperature are no longer correlated (Fukumura *et al.*, 1998). The dopaminergic pathways leading from the striatum to the anterior hypothalamus (Kiyohara *et al.*, 1984) are especially well positioned to relay a feed-forward signal to the hypothalamus to activate peripheral mechanisms before it receives feedback concerning the rise in core temperature.

A crucial role for dopaminergic mechanisms in heat-loss generation is suggested by a large body of evidence from animals and human clinical patients (see Previc, 1999). For example, reduced sweating and heat loss – sometimes reaching dangerous levels – are found in hypo-dopaminergic disorders such as Parkinson's disease and phenylketonuria and following use of anti-dopaminergic medication in schizophrenia (Figa-Talamanca and Gualandi, 1989). Also, anti-fever agents are known to block the pyrogenic actions of the prostaglandins, which normally inhibit dopamine (Schwarz *et al.*, 1982). There are many other neurophysiological mechanisms that contribute to thermal tolerance, but even these other mechanisms may be dependent on dopaminergic regulation. One such neuroendocrine mechanism is growth hormone, which is higher in heat-tolerant individuals (Niess *et al.*, 2003) but reduced in those with less plentiful eccrine sweat glands and sweat output (Hasan *et al.*, 2001; Lange *et al.*, 2001). Dopamine is a major stimulant for growth hormone, especially when growth hormone levels are low (Bansal *et al.*, 1981; Boyd *et al.*, 1970; Huseman *et al.*, 1986; Jacoby *et al.*, 1974), although it inhibits growth hormone production when it is excessive, as in pituitary adenomas (Bansal *et al.*, 1981).

As discussed in Chapter 2, expansion of dopaminergic systems represents the major neurochemical difference between primates and many other mammals such as rodents, a trend that evidently continued in humans. The precise mechanism through which dopamine levels became elevated is unclear, but the role of at least some genetic selection is suggested by the 1.4 percent genetic divergence of humans and apes and the co-occurrence of so many important changes that affected body shape and size as well as a host of physiological systems that increased thermal efficiency. The constellation of physiological changes described above need not have involved separate mutations for genes controlling each of these functions. For example, dopamine is known to increase growth hormone and in so doing contributes to elongating the body as well as promoting heat loss. An elongated body and greater thermal tolerance could also have led to greater bipedal locomotion and exercise, which in turn further increased dopamine levels by, among other actions, increasing calcium production and the calcium-dependent activity of tyrosine hydroyxlase, which converts tyrosine into dopa (Chaouloff, 1989; Gilbert, 1995; Heyes *et al.*, 1988; Sutoo and Akiyama, 1996). Similarly, the enhancement of bipedalism for running and heat loss ensured, through chase-hunting and scavenging, a greater supply of protein-rich meat sources containing rich amounts of tyrosine (Previc, 1999), which further increased the supply of dopamine (see Chapter 2). Increased protein from meat consumption is known to increase height (Suzuki, 1981)[7] and, because meat is easier to digest than raw plant food, a meat-based diet would permit the digestive tract to grow smaller (Henneberg, 1998), thereby reducing body mass relative to brain size and body surface and further aiding in heat dissipation.[8] It is not clear whether the other major physiological adaptation – namely, the reduction in hair growth – was due to a separate set of genetic mutations or associated with the above hormonal and neurochemical changes. Hair growth is controlled by a complex mixture of genetic and hormonal influences, and sometimes a given hormone may have the same hair-stimulating effect when in severe excess or deficiency (Alonso and Rosenfield, 2003). To date, no genetic factors have been found to underlie the differences between humans and chimpanzees in hair growth.

The consequences of bipedalism and the dopaminergic expansion may, in combination, account for most of the prominent brain features that distinguished *Homo habilis* and later humans from the Australopithecines. These include:

1. the aforementioned larger brain-to-body mass ratio, due to the effect of dopaminergically stimulated growth hormone to increase brain size relative to a stable or even decreasing digestive tract (see above);
2. a greater convolutedness of the cortex, due to the higher concentrations of dopamine in the upper cortical layers, which when enlarged relative to the lower layers tend to buckle and form fissures (Richman *et al.*, 1975); and

[7] For example, the height of the average Japanese was increased by a remarkable eight centimeters from 1949 to 1979 due to the increased consumption of meat (Suzuki, 1981).
[8] Possibly coinciding with the switch to a carnivorous diet, a genetic mutation specific to myosin and affecting jaw muscle strength and chewing strength has now been traced to the human lineage beginning about 2.5 million years ago (Stedman *et al.*, 2004).

3. lateralization of cortical function, most likely related to the switch to an upright stance that produced asymmetrical prenatal inertial forces during motion. The latter ultimately may be presumed to have led to a predominance of vestibular processing in the right hemisphere of most humans (Previc, 1991), a bias towards right-handedness, and the creation of a dopamine-rich left hemisphere that functionally bears little resemblance to the brain of the chimpanzee (Gazzaniga, 1983).[9]

As noted in Chapter 1, there is no evidence that the changes in brain size, convolutedness, and lateralization were in and of themselves important to the evolution of human intelligence and individually selected for genetically, so they may be considered mostly epiphenomena of the larger physiological/skeletal adaptation. Although one must always be cautious when only limited fossil evidence from the past is available, it is more plausible to infer from the dopaminergically enhanced physiology and diet of modern-day hunter-gatherers and the critical role of dopamine in advanced intelligence that it was the increase in dopamine that elevated the intellectual capacity of *Homo habilis* above its predecessors. Two of the many consequences of this increased physiological and intellectual capacity were the much more widespread and sophisticated tool use of *Homo habilis* – which gave rise to its Latin name meaning "domestic" – and its putative migration clear into Asia as early as 1.9 million years ago (Wanpo *et al.*, 1995).

More fundamentally, the increased distances traversed by *Homo habilis* in its scavenging and hunting would contribute to an increased emphasis on distant space and time – hallmarks of dopaminergic thought. But despite acquiring many modern-like features, the intellectual capacity of *Homo habilis* and its immediate successors was still quite inferior to that of modern humans. And, although the brains of *Homo erectus* and archaic *Homo sapiens* expanded greatly in absolute size, they grew much more modestly in relative brain-to-body size since humans also grew in height. Although they continued to migrate extensively throughout Eurasia, these protodopaminergic humans never achieved the intellectual level of their modern counterparts, as they used modifications of the same basic stone-flake tool design for hundreds of thousands of years. There is some evidence that slight improvements in blade technology did occur before 200,000 years ago (McBrearty and

---

[9] The belief that bipedalism contributed to cerebral lateralization in *Homo habilis* is shared by other theorists such as Corballis (1989) and is consistent with evidence of right-handedness (e.g. asymmetrical stone-flaking) in the *Homo habilis* archaeological record and an even greater dextrality in the *Homo erectus* record (Toth, 1985).

Brooks, 2000), and it is possible that the immediate descendants of *Homo erectus* may even have possessed a primitive language (Bickerton 1995; Corballis, 1992), perhaps even some capacity for speech (Kay *et al.*, 1998). Moreover, the modest but significant correlation between body height and intelligence in modern humans (Johnson, 1991) suggests that at least some of the dopamine-related physiological adaptations that elongated the body in the archaic humans may have been associated with increased intelligence. But despite the increasing physical resemblance of archaic humans to modern humans, the final leap forward to the modern human intellect required yet another confluence of events that occurred long after the ancestral genome and anatomy common to all modern humans were crystallized.

5.3   The emergence of the dopaminergic mind in later evolution

For at least 100,000 years after the emergence of a genome and anatomy common to all modern humans, the pace of human intellectual evolution continued to lag. This is somewhat remarkable in that humans clearly had the physical capacity for speech and language, but that capacity did not appear to be in and of itself sufficient to create the great intellect found in modern humans. The period from 100,000 to 200,000 years ago is part of the late Pleistocene geologic epoch and includes the beginnings of the Middle Stone Age, which most experts date at around 100,000 years ago but which has been pushed back by some to around 250,000–300,000 years ago (McBrearty and Brooks, 2000). While it is still debatable whether humans before 100,000 years ago had developed the beginnings of advanced blade and point technology, certainly the greatest technological achievements of the Middle Stone Age (shaped bone tools, barbed points, microliths, mining, fishing, beads, art, music, etc.) were all delayed until less than 100,000 years ago. Although a recent finding points to an isolated example of primitive bead production (shells with small holes) as far back as 90,000 years ago (Vanhaereny *et al.*, 2006), widespread and systematic advances in tool-making technology and art did not occur until around 70,000–80,000 years ago (Henshilwood *et al.*, 2001; McBrearty and Brooks, 2000), primarily along the coast of Southern Africa. To what extent many of these cultural advances appeared gradually has been the subject of intense debate. Mithen (1996) has been a leading advocate of a cultural "Big Bang" (although he argues it occurred as late as 50,000 years ago in Europe), whereas McBrearty and Brooks (2000) posit a more gradual advance (even though their own Figure 13 shows evidence of an abrupt jump in many technological areas around 80,000–100,000 years ago). Although some cultural achievements like musical instruments and animal artwork may date back to as little as 40,000–50,000 years ago (Fitch, 2006; McBrearty and Brooks, 2000), tally marks from 77,000 year-old artifacts found in Blombos Cave on the coast of South Africa suggest an earlier mathematical capability. An even more remarkable but still-controversial finding from Rhino Cave in Botswana suggests that animal sculpture and even worship occurred 70,000 years ago, marking the earliest evidence of religious ritual in modern humans (Handwerk, 2006). So, it would appear that not only was there indeed a "cultural Big Bang" but that it can be first traced to Southern Africa around 70,000–80,000 years ago. Moreover, recent genetic and other data now suggest that these humans quickly advanced from South Africa to the horn of Africa beginning around 70,000 years ago (see Figure 5.1). This relatively small population of humans, possibly after merging with a small population of East Africans, later crossed into the Middle East no later than 65,000 years ago and are believed to have become the ancestors of all present-day non-Africans as well as, because of back-migration, the vast majority of present-day Africans. The Khoisan of south-central Africa may be one of the few populations to contain remnants of the anatomically modern humans who first migrated from East Africa to South Africa 150,000 years ago (Behar *et al.*, 2008).

For those who firmly believe that human intelligence was the result of genetic adaptation, the >100,000 year gap between the emergence of genetically modern versus culturally modern humans is highly problematic, even if one accepts the youngest limit for the common ancestor from most genetic datings (150,000 years ago) (Hammer, 1995). And if older modern artifacts are eventually found, this would not negate the fact that the overwhelming majority of human artifacts as recently as 100,000 years ago fail to manifest the advanced characteristics that became widespread *just 30,000 years later*. It is hard to imagine the majority of present-day humans exhibiting so little a tendency to create art or fishing barbs or bone tools for that long. All of these arguments support an *epigenetic* explanation for the explosion of culture within the past 100,000 years. This would be consistent with the extreme unlikelihood that there was a direct genetic selection for any specific cognitive abilities required to support these cultural achievements (see Wynn and Coolidge, 2004; Chapter 1, this volume). Indeed, if cognitive selection were of such great adaptive value, why did it take over one million years from the appearance of *Homo erectus* to create the first modern human artifacts and, even then, only in a single region of the world (i.e. Southern Africa)?

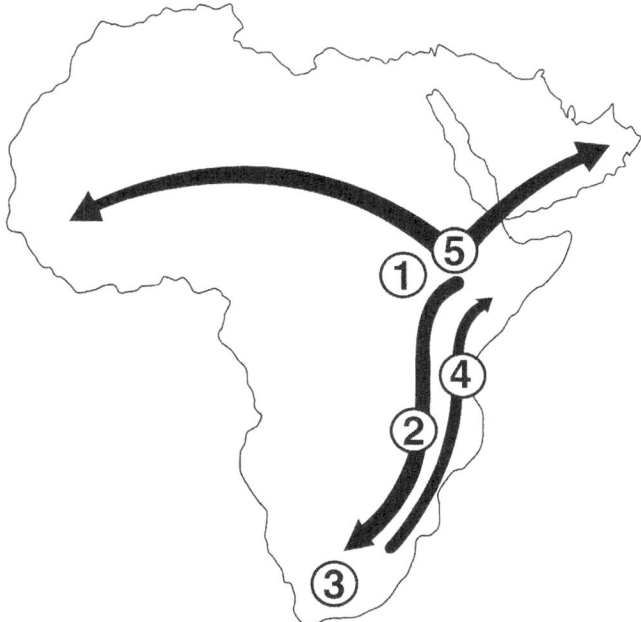

Figure 5.1 The hypothesized direction of modern human origins and migration.

Notes and sources:
1. modern humans first emerged in East Africa around 200,000 years ago (McDougall *et al.*, 2005);
2. part of the East African population migrated to Southern Africa beginning around 150,000 years ago (Behar *et al.*, 2008);
3. "modern" human intelligence occurs along the coast of South Africa around 70,000–80,000 years ago (Henshilwood *et al.*, 2001);
4. migration of one portion of the South African population back to East Africa occurs shortly thereafter (Behar *et al.*, 2008);
5. the merged East African and South African populations migrate along the South Asian coast all the way to Australia, beginning around 65,000–70,000 years (Macaulay *et al.*, 2005), as well as to other parts of Africa (Behar *et al.*, 2008).

One plausible explanation for the intellectual advances in Southern Africa during the late Pleistocene era is the change in diet – specifically, the widespread consumption of shellfish – that occurred around 100,000 years ago in coastal areas of South Africa and possibly elsewhere. A second, more speculative explanation relates to demographic pressures that increased population density and cultural exchange. As reviewed in the next sections, both of these changes could have greatly elevated brain dopamine to a level capable of supporting the newly advanced intellect and, through epigenetic inheritance, be passed on to subsequent generations of humans.

5.3.1   The importance of shellfish consumption

Some researchers (e.g. Erlandscn, 2001) believe that human consumption of marine fauna began 150,000 years ago, possibly due to a coastal migration to escape increasingly arid inland regions, and isolated evidence for possible shellfish consumption has been found in Eritrea dating back 125,000 years (Walter et al., 2000) and at Pinnacle Point in South Africa around 160,000 years ago (Marean et al., 2007). However, it is generally accepted that *widespread* transport and consumption of shellfish occurred less than 100,000 years ago along the South African coastline (Broadhurst et al., 1998; Dobson, 1998; McBrearty and Brooks, 2000). Shellfish are rich in iodine and essential fatty acids (Broadhurst et al., 1998; Dobson, 1998), both of which have been shown to increase dopamine activity and intellectual development (DeLange, 2000; Wainwright, 2002). Iodine, found in red meat but especially in marine animals, is converted by the thyroid gland into thyroxine, which in turn stimulates tyrosine hydroxylase and the production of dopa (see Previc, 2002), while essential fatty acids are known to increase dopamine receptor binding and dopamine levels (Delion et al., 1996; Wainwright, 2002). As noted earlier, inadequate iodine, leading to reduced levels of dopamine and norepinephrine, is the single largest preventable source of mental retardation in the world today (Boyages and Halpern, 1993; DeLange, 2000). Conversely, hyperthyroidism is frequently associated with dopaminergically mediated psychosis (Benvenga et al., 2003) More generally, the timing and amount of thyroid output is an important influence on mammalian speciation, and it has been specifically implicated in the process of domestication (Crockford, 2002).

As reviewed by Gagneux et al. (2001) and Previc (2002), there appears to have been a major increase in thyroid hormone production (particularly of T3, which is formed from iodine) with the advent of modern humans. This finding is consistent with the similarity of many superficial physical features of chimpanzees and those of iodine-deficient humans (O'Donovan, 1996) and the relatively enlarged human thyroid gland relative to that of the chimpanzee (Crile, 1934). By contrast, chimpanzees and other mammals have relatively larger adrenal glands, which are important for transient arousal, which brain dopamine inhibits. One important difference between Neanderthals and modern humans appears to have been the amount of iodine in their bodies, since Neanderthals exhibit many skeletal features (large femurs, short stature, extended brows, larger brains) characteristic of iodine-deficient modern humans (Dobson, 1998) and because prehistoric modern humans are believed to have consumed relatively more shellfish than big game (Dobson, 1998; McBrearty and Brooks, 2000; Richards *et al.*, 2001). The large amount of shellfish consumed is especially noteworthy in Klasies River excavations, but our modern human ancestors generally tended to inhabit coastal areas, rivers, and large lakes where aquatic fauna would be easy to obtain. There is some tentative evidence that Neanderthals, at least early on, were less likely to inhabit coastal or otherwise iodine-rich regions than modern humans (Dobson, 1998), even as later Neanderthals began to consume more marine fauna. It has been long debated why modern humans so quickly replaced Neanderthals in Europe after 50,000 years ago, but population replacement has been widespread among even among modern human populations and the entering population usually possesses greater dietary breadth in these instances (O'Connell, 2007). The Neanderthals' over-reliance on large game animals may have been particularly disastrous when the last Ice Age approached its peak around 30,000 years ago as large game became less plentiful and subject to competition with growing populations of modern humans. The relative inability of Neanderthals to successfully switch to more diverse sources of food may not have had anything to do with their brains, in that modern human populations frequently show resistance to changing their food procurement strategies as their territory is encroached upon by other human groups, which can mainly be attributed to socio-cultural factors (O'Connell, 2007).

Given the importance of the thyroid gland in human brain development and function, it is reasonable to conclude that the consumption of shellfish was a major impetus for bringing the dopaminergic mind to fruition. Indeed, advanced tool-making artifacts in the Middle Stone Age coincide both geographically and temporally with the consumption of shellfish at various coastal sites in Eastern and Southern Africa (McBrearty and Brooks, 2000; Walter *et al.*, 2000). However, increased shellfish consumption alone could not have led to advanced intelligence, because crab-eating macaque monkeys would then rank among the most intelligent of species. It may be presumed that the later dietary-induced rise in dopamine in humans proved so effective partly because it was built on the earlier increase in dopamine that was part of the general physiological adaptation for endurance activity. Moreover, the increase in shellfish consumption was probably not the only nongenetic factor responsible for the dramatic cultural advance beginning less than 100,000 years ago, because even current iodine-deficient humans, despite their reduced intellect, are capable of art and religion. Moreover, since humans probably began to consume marine fauna in large quantities at least several millennia before the explosion of advanced intellectual artifacts around 80,000 years ago, shellfish consumption may have contributed to but not guaranteed the emergence of the dopaminergic mind.

5.3.2   The role of population pressures and cultural exchange

Although directly contributing to improved intellectual functioning, the evolution of human intelligence and behavior was aided by a more general benefit of consuming marine fauna – namely, that it greatly improved and stabilized the human diet and thereby *promoted a longer lifespan* (McBrearty and Brooks, 2000). The increased longevity, in turn, is believed to have led to population pressures, increased migration, and increased exchanges among small groups of humans (McBrearty and Brooks, 2000; Mellars, 2006). Although it has been argued that genetically modern humans experienced a major population decline due to drought and other climatic conditions (Ambrose, 1998), this claim has been refuted (Hawks *et al.*, 2000) and was unlikely to be true in South Africa, which had a plentiful supply of marine fauna. Certainly, the remarkable number of major archaeological sites from Cape Town on the west to Port Elizabeth on the east (Henshilwood *et al.*, 2001) suggests that South African coastal regions were well-populated in the Middle Stone Age. And, skeletal evidence indicates that modern humans in Europe had a slightly longer lifespan than their Neanderthal cousins (Arsuaga, 2003). Longevity-induced population pressures would presumably have followed the rise of shellfish consumption with a lag, and environmental changes in sub-Saharan Africa between 70,000 and 80,000 years ago are also believed to have further increased the competition for resources and sped up migration and interchange among populations (Mellars, 2006). These twin factors could explain the lag between the initial widespread consumption of shellfish and the remarkable cultural advances 75,000 years ago.

Increased social exchange is known to be a catalyst for intellectual advances in other advanced primates, such as Sumatran orangutans (Van Schaik, 2006), and the role of the cities in promoting social exchange and intellectual progress has long been recognized.[10]

---

[10] A fascinating recent example of the role of social factors in cognitive development is that of deaf Nicaraguan children who had previously lived at home in linguistic isolation

Moreover, increasing population growth and exchanges could make the identification with one's group even more important. The growth of art and possibly music in sub-Saharan Africa toward the end of the Middle Stone Age years ago may both be manifestations of this trend, given their roles in social identification and group cohesion (Brown, 2000; Shea, 2003). Moreover, the slower cultural advancement in South Asia relative to Europe toward the end of the Middle Stone Age (60,000–30,000 years ago) as well as the slower cultural progression of Neanderthals relative to modern humans have both been partly attributed to the higher population densities of modern humans in Europe (James and Petraglia, 2005; Wynn and Coolidge, 2004).

Increased population and increased social and cultural exchanges may not only have benefited from increasing intellectual prowess but may, in turn, have contributed to them biologically. Cultural evolution is known to alter biological inheritance (Laland et al., 1999), and the competitive stresses and achievement drives of modern urban society are believed to contribute to elevated dopamine levels (Pani, 2000; Previc, 2007). Indeed, hyperdopaminergic disorders such as schizophrenia and autism are more likely to occur in urban areas (Lauritsen et al., 2005; Marcellis et al., 1999; Palmer et al., 2006), and a milder form of these pressures could have been present to a certain extent even in sub-Saharan Africa in the late Pleistocene. On the other hand, it is difficult to imagine a current human even in a sparsely populated region not producing art, ornaments, or other advanced objects, and certainly no densely populated primate population has ever managed to even approach the intellectual output of the average human. Nevertheless, increased dopamine due to dietary improvements and increased social interaction, building upon an already protodopaminergic mind, may have collectively been sufficient to lay the foundations of the true dopaminergic mind.

In line with the relatively stable genome, no major anatomical change was associated with this later dopaminergic progression, although the human brain did slightly decrease in size from ~1,500 cc to the present-day 1,350 cc (Carroll, 2003). But, once hominid evolution had progressed to its later stage, genetic changes were no longer necessary for the dopaminergic increase to be expressed throughout the entire species. Cultural and dietary influences on dopamine, transmitted prenatally, would have been passed on and enhanced in successive generations and thereby effectively been rendered a permanent part of our inheritance. Even when humans moved inland and no longer relied as much on

and then were placed in a group setting and readily formed a new (emergent) language on their own (Senghas et al., 2004).

aquatic fauna, their dopaminergically mediated cultural achievements were self-sustaining. Thus, not only was the relatively stable human genome and anatomy beginning around 200,000 years ago insufficient in catalyzing the intellectual explosion that occurred over 100,000 years later, it was also *unnecessary* to maintain the newly advanced intellect as modern humans left Africa and spread throughout the world.

5.4   Summary

According to the scenario just presented, the dramatic expansion and increasing functional importance of dopamine systems in the human brain occurred over a long period of time, only partially paralleled by the changes in our physical appearance, and the dopamine rise continued long after our common genetic lineage. For those who believe that the great intelligence of humans relative to the rest of the primate world requires an exalted evolutionary progression, this theory is highly disappointing in that:

1. there is no positing of specific genetic mutations for language or other advanced intellectual functions;
2. no importance is ascribed to the changes in cranial size and shape; and
3. most of our intelligence is regarded as a by-product of physiological adaptation, diet and population pressures.

This theory is, however, consistent with a number of crucial facts, including:

1. the lack of a causal relationship between brain size and intelligence;
2. the tenuous relevance of the human genome to intelligence;
3. the importance of dopamine to our advanced intellect (Chapter 3);
4. the specific expansion of the dopamine-rich striatum and cerebral cortex relative to the rest of the brain in humans as compared to chimpanzees (Rapoport, 1990);
5. the known effects of diet and physiology on dopamine function;
6. the sub-Saharan origins of humanity; and
7. the >100,000 year gap between the establishment of the modern human anatomy and genome and the appearance of a variety of cultural artifacts associated with a distinctly modern human intellect.

This theory merges genetic, epigenetic and cultural factors and blends the African gradualistic theory of McBrearty and Brooks (2000) with the general intellectual explosion posited by Mithen (1996), but with his European locus of the intellectual "Big Bang" now placed at an earlier juncture in Southern Africa. In addition, this theory clearly explains why cognitive abilities like language, religion, art, advanced tool-making, mining, long-distance exchange etc. did not emerge as separate genetic selections but rather depended on a more fundamental increase in cognitive potential as expressed in enhanced working memory, cognitive flexibility, spatial and temporal distance (off-line thinking), mental speed, and creativity that could only be found in a brain rich in dopamine (see also Wynn and Coolidge, 2004). Finally, this theory dispels the notion that the evolution of advanced human intelligence was critically dependent on the capacity for speech, since the latter was present at least 200,000 years ago in the first anatomically modern humans and probably another 100,000 or 200,000 years before that in archaic humans (Arensburg *et al.*, 1990), without leading to any dramatic intellectual advances.

By highlighting the epigenetic inheritance of dopaminergic activity, this scenario shows why the emergence of the dopaminergic mind was not associated with a unique, immutable genetic process but rather one that continued to advance long past the establishment of our common genome and anatomy. In fact, as described in the next chapter, the dopaminergic mind appears, at least in the industrialized nations, to have undergone at least two additional transformations during our more recent history.

6   The dopaminergic mind in history

If one believes that human evolution – especially in its intellectual aspects – did not rely exclusively or even largely on brain size and genetic transmission, then human evolution has never ceased. Hence, it would be incorrect to assume that all genetically modern humans and societies have the same neurochemistry and therefore the same intellectual abilities, personalities, goals, and propensity toward mental disorder. In particular, there is reason to believe that levels of dopamine are now much higher in members of modern industrialized societies than in more primitive societies. This chapter will focus on two major historical epochs – the transition from the hunter-gatherer societies to the ancient civilizations and the dramatic expansion of the dopaminergic consciousness and lifestyle in the twentieth century. In so doing, this chapter will highlight the role of influential individuals in history who have manifested dopaminergic traits and behaviors and played important roles in shaping our modern dopaminergic world.

6.1   The transition to the dopaminergic society

Despite a certain degree of technical proficiency, Neanderthals and even modern humans for their first 100,000 years appear to have lacked the generativity and pervasive "off-line" thinking capabilities of later humans. Once the prehistoric cultural "Big Bang" had progressed to its final stages and the last great Ice Age began to recede around 20,000 years ago, intellectual evolution proceeded at a rapid pace, with seemingly but an instant required from the Neolithic Era and the beginnings of agriculture to the ancient civilizations and the Copper, Bronze, and Iron Ages. Intermixed with occasional periods of stagnation and despair, cultural advances in the form of literature, mathematics, legal codes, navigation, engineering, and architecture grew steadily over many millennia and in many places around the world, often independently of each other. Eventually, the gradual accumulation of knowledge and commerce gave way to the European Renaissance, which set the stage for the final expansion of the dopaminergic mind. However, it was not until the second half of the twentieth century that the dopaminergic mind reached its ultimate state. Evidence of this heightened twentieth-century dopaminergic state is found in rapid increases in dopaminergically mediated intelligence and technological advances (Dickens and Flynn, 2001) and in a host of hyperdopaminergic mental disorders. Unlike the earlier increases in dopamine levels, there were no major environmental adaptations or dietary changes associated with the dopaminergic rise since prehistoric times. Rather, most of the dopamine increase occurred because higher dopamine levels became highly adaptive to individuals surviving and prospering in the increasingly stratified, complex, and competitive socio-economic systems that emerged during human history.

It may be instructive to review briefly how different the earliest hunter-gatherer societies must have been relative to our current industrial and post-industrial societies. It has been repeatedly documented that present-day hunter-gatherers work at most only about 20–30 hours a week, with little time lost for commuting. Although some authors claim the hunter-gatherers comprise what Sahlin termed "the original affluent society" (Sahlin, 1972; Taylor, 2005), their amount of leisure time has probably been somewhat exaggerated, at least in the case of females (Caldwell and Caldwell, 2003; Hawkes and O'Connell, 1981). Nevertheless, rather than compulsively arising at a certain time to begin each workday – or constantly worrying about the future – work appears to be more sporadically engaged in and there is a much greater "present orientation." Work is also engaged in as a communal activity, sharing of resources is the norm (Taylor, 2005), and most leisure time is also spent in communal activities such as games and music (Sahlin, 1972), the latter of which is believed to have emerged mainly as an activity to promote socio-emotional bonding (Brown, 2000). Even agrarian life is considered too demanding by many hunter-gatherers, who given the option prefer the simpler hunter-gatherer life that is mostly devoid of material possessions (Sahlin, 1972). Indeed, material rewards unrelated to immediate consumption (i.e. secondary rewards) are of limited value in these societies, since they only create logistical problems given the frequent moves to new locations required by the nomadic existence. The per capita energy consumption of the hunter-gatherer is estimated at 3,000 calories per day, or less than 1 percent of modern humans, who have extensive transportation and electrical power requirements (Clark, 1989: 102). And, hunter-gatherers are largely free of the diseases and epidemics associated with urban existence, such as measles, influenza, plague, smallpox etc. Of course, there would have been many diseases unique to the prehistoric hunter-gatherers, and their overall lifespan has been estimated at only twenty-one years (Acsadi and Nemeskeri, cited in Caldwell and Caldwell, 2003), which is considerably less than that of current hunter-gatherers (~30–35 years), who have much better tools and engage in limited cultivation (Arsuaga, 2003).

Taylor (2005) and others have reviewed evidence that early hunter-gatherer societies were relatively nonviolent and highly sexually permissive, consistent with the high rate of sexually transmitted disease among current hunter-gatherers such as the !Kung San (Caldwell and Caldwell, 2003). Certainly, there were no standing armies among the hunter-gatherers, as they lived in small bands, and many of them were undoubtedly very peaceful, based on their artwork, their skeletal remains, and the fact that many present-day hunter-gatherer societies are basically nonviolent (Taylor, 2005). But, there is certainly evidence of violent confrontations involving many prehistoric groups, such as the Caribs and Yanomamo of South America, and up to 25 percent of prehistoric adult male deaths have been ascribed to violence (Coleman, cited in Caldwell and Caldwell, 2003). It is generally agreed that prehistoric hunter-gatherer societies were much more egalitarian than were the agrarian societies that followed, with women and men sharing child-rearing and work, and most of these societies – as do a large number of remaining ones – presumably had a maternal lineage system. Their religious systems mostly did not emphasize remote, otherworldly, or high gods (e.g. "sky gods") but rather invoked natural and ancestral spirits. Finally, the prehistoric peoples tended to live more within nature than did the agrarian descendants. This does not mean that they were environmental angels – after all, many if not most of the large mammals and marsupials outside of Africa and Asia disappeared before the dawn of agriculture, and the burning practices of prehistoric societies contributed to deforestation and erosion in many parts of the world (Roszak, cited in Taylor, 2005).

As the Earth began to warm substantially around 20,000 years ago, the hunter-gatherer lifestyle eventually gave way to the development of primitive cultivation (e.g. seeding of wild grains and fruit trees) and nomadic herding that probably began 10,000–15,000 years ago as population pressures reduced the availability of large game animals in Southwest Asia (Arsuaga, 2003; Clark, 1989; Weiss and Bradley, 2001). The mixture of hunting and gathering, migratory grazing and farming was eventually replaced by a much greater reliance on cultivation of the soil and the development of permanent agrarian settlements around 10,000 years ago, which led to an increase in energy consumption to ~15,000 calories per day. It is widely believed that the agrarian lifespan may have initially slightly decreased (Clark, 1989; Taylor, 2005), as more confined settlement living exposed humans to more epidemics and animal diseases. However, there is also evidence that human populations may have expanded more rapidly than before, which indicates that an improved diet may have increased fertility and offset a higher mortality (Caldwell and Caldwell, 2003).

With the advent of agriculture, the psychological relationship with nature was dramatically altered – instead of living with nature, the agrarian societies were dedicated to transforming (controlling) nature through cultivation, irrigation, harvesting etc. Unlike the hunter-gatherers, who had few permanent possessions and could easily switch diets in times of drought and/or migrate, the agrarian societies were based on an economy and lifestyle that was more sedentary and cyclical, depending on storage (hoarding) and prediction of seasonal weather for planting and harvesting (Clark, 1989). Both hoarding and prediction require a greater orientation to future events, which is a hallmark of the dopaminergic personality (see Chapter 3); indeed, hoarding is a classic obsessive-compulsive symptom and is dependent on the integrity of dopaminergic systems (Kalsbeek et al., 1988). The greater sensitivity to lunar and solar cycles also led to an expanded sense of dopaminergically mediated upper space, and otherworldly religious symbols (i.e. sky gods) began to replace the mythical animal and human figures that constituted the first religious icons.

The agrarian existence led to the emergence of several "proto-civilizations" in what Taylor (2005) refers to as Saharasia, a broad stretch of territory extending from North Africa to the Indus River. During the first several thousand years after the beginnings of agriculture, the warming earth and melting glaciers produced rainfall in this broad region that was mostly sufficient to support vast savannas and even forests at the higher elevations. Food was plentiful, the settlements were not all that densely populated – probably only a hundred to a few thousand persons in the larger ones – and the societies were generally peaceful and egalitarian (most homes were of the same basic type and size). Earth goddesses and women priests were prominent in the early civilizations, and sexual behavior was still relatively permissive.

As recounted by Taylor (2005), however, global temperatures continued to rise and the last remaining glaciers in the region began to melt. Along with overgrazing and other rapacious agricultural practices, the global climate change led large swaths of Saharasia to dry out, and increasing competition for resources ensued. By 6,000 years ago, cities had grown larger, social stratification increased, warfare suddenly erupted on a large scale, and masculine values began to predominate.

The constant warring that unfolded was more intellectualized, senseless, and disturbing than in prehistoric times, as it was often a meglomaniacal display of power by rulers and was frequently accompanied by mutilation and torture (Wilson, 1985). The natural religions were replaced by otherworldly hierarchical religions in which mostly male gods and male priests reigned supreme, and a different attitude toward nature emerged (dominance rather than co-existence). Technological achievements in the Bronze and Iron Ages occurred at an astonishing pace, fueled partly (or perhaps even largely) by the need for better weapons and other armor; indeed, so great was the technological prowess of the ancient civilizations that some of their products – notably, the famous Antikythera Mechanism, which was an astronomical calculator made of thirty geared wheels – were more sophisticated than anything produced for the next 1,000 years (Charette, 2006). Daily life became much more "propositional" – i.e. regulated by complex, abstract laws and written communications. A more negative attitude toward nature, maternal symbols of nature, and the body itself emerged, and sexual permissiveness and partial nudity were replaced by body coverage and a multitude of sexual prohibitions. The glorified status of male warriors and the concentration of wealth in much more stratified societies ultimately led to the replacement of matrilineal descent and inheritance by patrilineal descent and inheritance, which may have further restricted the sexual freedom of females.[1] The populations of Central and Southwest Asia may have been among the first to exhibit these new attitudes and lifestyles, but other civilizations soon followed and large fortressed city-states were constructed throughout this region and even in parts of Europe, Africa and East Asia. The most wrenching part of the environmental change occurred just after the end of the dramatic post-glacial sea rise known as the Flandrian Transgression, during which a worldwide drought of almost unimaginable proportions between 2300 and 2000 BC led to the collapse of many civilizations around the world and the emergence of large refugee populations (Bowen, 2005; Weiss and Bradley, 2001). By 1500 BC, one of the last of the relatively egalitarian, peaceful and permissive early Saharasian civilizations (the Minoan in Crete) was conquered by the more aggressive Myceneans.

---

[1] As reviewed by Primavesi (1991), the co-emergence of masculine dominance and property wealth created a need for a strict patriarchal lineage that necessarily restricted female sexual freedom. Unlike in the matriarchal society, where lineage is obvious from birth, patriarchal lineage can only be presumed if female sexual behavior is limited exclusively to the husband.

The dramatic transformation of human existence into a much more competitive, inegalitarian, and ruthlessly technological one has been termed "The Fall" by Taylor (2005), and it may be the basis for the biblical story of Adam and Eve.[2] It has also been associated with a major change in human consciousness. Humans before this era viewed themselves as surrounded by and controlled and even aided by various gods; indeed, what might be today considered pathological symptoms such as hallucinations and delusions of being controlled (the "alien-control" syndrome) may actually have been quite common and tolerated during this time (Jaynes, 1976). The new consciousness, however, posited humans as independent, self-conscious agents, and the gods that they had once placed such faith in but who failed to prevent several centuries of natural catastrophes became more distant and less involved in controlling their thoughts and actions (Jaynes, 1976). Many theorists, including Jaynes (1976), Taylor (2005), and Wilson (1985), have argued that this era represents the beginning of left-hemispheric dominance, inner thoughts, ego explosion, masculine aggression, other-worldliness, a linear temporal perspective, and a technological mindset. It is certainly true that these features are more characteristic of the left hemisphere, but there is no actual evidence that our anatomical or functional lateralization was altered during this epoch. Rather, what occurred was a dramatic increase in the *dopaminergic mind*, which is typified by its advanced cognition, competitive drive, masculine style, focus on distant (future) goals and space, and control over nature and others ("agentic extraversion"), all at the expense of empathy towards others and maintenance of the emotional self. The dopaminergic mind merely masqueraded as an increased left-hemispheric dominance because the higher dopamine content of the left hemisphere makes it most typical of that mind in general. Furthermore, the lateral dopaminergic system may have been the part of the dopaminergic brain/mind that increased the most, because the internal control, ego-strength, techno-logical abstraction, and diminution of outside thoughts (hallucinations)

---

[2] Just as the first great civilization (the Sumerian) emerged in Mesopotamia (now Southern Iraq), so can the Garden of Eden be traced to this same general location (Hamblin, 1987) and timeframe (4000–5000 BC). In Eden, Adam and Eve (who was supposedly formed from the rib of Adam according to Sumerian myth) lived in a lush forest (which was present in many parts of the Middle East 6,000–8,000 years ago) and lived without shame of their bodies (sexual permissiveness), and in a state of techno-logical ignorance. But, after disobeying God, Adam and Eve acquired knowledge, shame of their bodies, and were kicked out of lush Eden and into the desert (which started expanding around this same time), and for the first time greed and violence occurred in their offspring.

that all began to dominate human consciousness at this time are most characteristic of the lateral system (see Chapter 3).[3]

Unlike the previous expansions of the dopaminergic mind, this one was not due to a genetic evolution, physiological adaptation or transformed diet. Although brain size in genetically modern humans did decrease slightly to 1,350 cc, it had stabilized before the rise of the proto-civilizations, and the genetic variability among extant humans is remarkably low, only ~0.1 percent (Carroll, 2003). Moreover, the fact that patriarchal, stratified, sexually restrictive, warlike societies with sky-gods, priest classes, and a high level technological capability occurred in other civilizations throughout the world (e.g. the Aztecs in Mexico, the Mayans in Central America, the Incas in South America, and the Chinese in East Asia) suggests that it was principally *the emergence of competitive and stratified societies that forced the elevation of dopamine in its individual members.*

The late-Neolithic change in human consciousness associated with the rise of the first dopaminergic societies would not be the last time in history that humans themselves would change their neurochemistry and pass it on to their offspring. Led by men with highly dopaminergic minds, the human race continued to expand in its technological and scientific sophistication in all parts of the world – and particularly in Europe, the Middle East, and East Asia. Beginning at the time of the Renaissance in Europe and the age of the great explorers and accelerating into the twentieth century, the dopaminergic mind entered a new stage and eventually came to its full (and arguably overripe) fruition.

In the next section, I will review how dopaminergic traits in some of the leading (albeit flawed) figures in modern history led them to great accomplishments that would help shape the modern world, before concluding with a brief discussion of the modern hyperdopaminergic society.

---

[3] A central thesis of Jaynes (1976), based on an analysis of the cultural and literary record from the ancient civilizations, is that between 2500 BC and 1500 BC the inner thoughts of humans that were once attributed to external voices (e.g. the gods), in much the same way as are the hallucinations of people with schizophrenia, began to be more attributed to the voices of inner consciousness. In addition, evidenced in Homer's *Odyssey*, humans increasingly began to be portrayed as in possession of their own thoughts and actions, and there were more literary references with abstract meaning (e.g. nouns ending with "ness") and specific temporal connotation (e.g. "hesitate"). As noted in Chapter 4, there is substantial evidence that the inhibitory control of the lateral dopaminergic system over the subcortical portions of the medial dopaminergic system is weakened in schizophrenia, thereby leading left-hemispheric inner speech to be perceived as auditory hallucinations.

6.2   The role of dopaminergic personalities in human history

The evolution from the hunter-gatherer and agrarian societies to our current industrialized societies occurred in but a speck of geological and even biological time. However, it did not occur overnight, and it did not necessarily occur by accident. Rather, human societies during the past 4,000 years have almost always been led or influenced by dominant individuals in the form of military rulers, religious leaders, explorers, and scientists. Highly dopaminergic individuals may be among the 2 percent or so of high-achievement individuals described by Toffler (1970), who are seemingly well-adapted to the environments of modern societies and who are in most cases its leaders. However, even highly successful individuals may be prone to hyperdopaminergic syndromes such as hypomania (Goodwin and Jamison, 1990) and obsessive-compulsive spectrum disorders such as workaholism, excessive risk-taking, and sexual addiction, and these individuals also may be part of families in which hyperdopaminergic disorders are common (Karlsson, 1974). It is also interesting to note in this regard that famous persons are, like those who suffer from schizophrenia and bipolar disorder, more likely to be born in the winter months (Kaulins, 1979), possibly due to high maternal dopamine levels in the summer months caused by higher temperatures and longer daylight hours (see Chapter 4).[4]

The fact that almost all famous individuals in history prior to the twentieth century were male is consistent with Carlyle's famous quotation that "The history of the world is but the biographies of great men."[5] This phenomenon at least partly reflects the general predominance and admiration of male traits in human societies since the Bronze Age. But, most famous historical figures were not only male but also very youthful (<thirty years old) at the time of their greatest accomplishments.[6] In an extensive treatment of "greatness," Simonton (1994) showed that

---

[4] Kaulins (1979) recorded the birth dates of 11,439 famous people listed in the 1974 edition of *Encyclopedia Britannica*. Across all fields, there was an 18 percent increase in January and February births relative to births in July and August. Of thirty-one fields categorized by Kaulins, twenty-three showed the excess winter-birth trend. (Kaulins actually had thirty-three categories, but two of them had only one person in each season.) See www.lexiline.com/lexiline/lexi118.htm.

[5] www.brainyquote.com/quotes/quotes/t/thomascarl143133.htm.

[6] Of the five individuals to be reviewed in this chapter, the only exception to this was Columbus, who was around forty-one at the time of his first voyage to the Western Hemisphere. However, his first formal petition to sail westward to the Indies was made eleven years earlier, and his dreams of sailing westward to glory probably began long before that.

geniuses are generally most productive around age forty, although this depends somewhat on:

1. the particular profession, in that mathematicians and physicists and artists tend to peak younger than medical and social scientists;
2. average lifespan, since until the twentieth century the age of death for surviving adults was less than fifty-fve years, which favored output at a younger average age; and
3. the maturity of a field of endeavor.

The last factor may be increasing the peak age of productivity in modern societies, since at least ten years of experience are required in most fields today to do one's best work (Simonton, 1994) and since most scientists today have so much to learn that they are still engaged in post-doctoral research even into their early thirties. On the other hand, many of the creative breakthroughs that led to significant achievements later on in life are known to have occurred much earlier in the person's life, so the age of peak creativity is undoubtedly younger than the age of peak productivity.

It has recently been argued that, for better (e.g. science) or worse (e.g. criminal activity), great male achievements occur early in adulthood because of high testosterone levels and the need of males to attract desirable female partners (Kanazawa, 2003). There may, indeed, be a testosterone link – as well as more mundane causes like greater family and other responsibilities, poorer health, and decreased energy levels that reduce productivity in middle age – but the testosterone link may not be as direct as implied. Testosterone increases the male sex drive, but it also elevates brain dopamine levels, which in turn increase aggressiveness, obsessiveness, and male sexual behavior. As previously discussed in Chapter 4, the testosterone–dopamine link accounts for why males are more likely to develop such hyperdopaminergic disorders as obsessive-compulsive disorder, mania, and schizophrenia in early adulthood. The decline in creativity and achievement in males with age is, therefore, highly consistent with the decline in dopamine levels in aging (Braver and Barch, 2002; Volkow et al., 1998).

Great historical figures cannot merely be considered obsessed, aggressive, and hypersexual, though many of them were clearly sexually promiscuous (Alias, 2000), including evidently a surprising number of great physicists such as Albert Einstein and Richard Feynman. What most of these famous men did possess was great intelligence, a superior working memory capacity, and a keen strategic vision (i.e. "far-sightedness"), in line with the dopaminergic emphasis on distant space and future time. These intellectual traits were coupled with a sense of personal (even religious) destiny and a restlessless and ambition to discover or conquer new worlds and/or promote new ideas. On the negative side, many if not most of these men were emotionally cold, somewhat deluded, and prone to other dopaminergic traits such as excessive risk-taking.

There are dozens of highly dopaminergic men of historical significance who could have been spotlighted in this chapter for their enormous working memories, incredible mental focus, and strong goal-directedness (see Alias, 2000; Simonton, 1994). The only persons to be highlighted here will be Westerners, partly due to my bias. I neglect to discuss two of the most infamous leaders of modern history (Adolf Hitler and Josef Stalin), not because they were lacking in dopamine – their clinical tendencies toward bipolar disorder and paranoia speak to the contrary – but because their negative "accomplishments" (arguably the greatest mass murderers in history) dwarf their positive ones. Nor have I highlighted any religious figures, again not because they were devoid of highly dopaminergic brains – on the contrary, religious drives and experiences are very much associated with high levels of dopamine (Previc, 2006) – but rather because their contributions are less amenable to objective historical analysis due to their perceived religious stature and, in many cases, a lack of contemporary and historically valid accounts. To illustrate the role of the dopaminergic mind in history, I have chosen to review briefly five persons who were arguably among the most influential in history – Alexander the Great, Christopher Columbus, Isaac Newton, Napoleon Bonaparte, and Albert Einstein (see Figure 6.1). Two of these rank among the greatest of military leaders, one was arguably the greatest explorer in history, and the other two arguably rank as the two most famous scientists ever. On any list of historical greatness, all would rank in the upper echelons because of their achievements. Though their personalities were different in many respects, they all shared a high degree of intelligence, a sense of personal destiny, a religious/cosmic preoccupation, an enormous focus (obsession) with achieving supreme goals and conquests, an emotional detachment that in most cases led to ruthlessness, and a risk-taking mentality that led to consequences ranging from the merely embarrassing to the outright disastrous.[7]

---

[7] Some of the "dopaminergic" also-rans include such intelligent, complex, and morally ambiguous (e.g. idealistic and ruthless) figures as Julius Caesar, Mustafa Kemal (Ataturk), and Winston Churchill. Another fascinating and influential dopaminergic mind – Howard Hughes – was the subject of the recent Hollywood blockbuster *The Aviator*. Hughes was a brilliant, driven, and creative aviator and movie pioneer who became the world's first authentic billionaire, but he was also wracked by debilitating hyperdopaminergic symptoms as obsessive-compulsive disorder, tics (a feature of Tourette's syndrome), and delusional and paranoid thoughts (schizotypy).

Figure 6.1 Five famous dopaminergic minds in history: Alexander the Great, Christopher Columbus, Isaac Newton, Napoleon Bonaparte, and Albert Einstein.

Public domain images courtesy of the Library of Congress.

6.2.1   Alexander the Great

The Macedonian king Alexander (356–323 BC) ranks, alongside Ghengis Khan, as one of the two greatest conquerors in history. But, like Ghengis Khan, his greatness stems not just from his military victories but also from his great intellect and vision for the administration of his empire, most of which did not come to fruition because of his early death. Unfortunately, Alexander's legacy was also marred by a ruthlessness that, while not approaching that of the Mongol leader, nevertheless is shocking by today's standards.[8]

Alexander was raised as the son of Philip, king of Macedonia. He had a keen intellect, and one of his teachers was Aristotle. Upon his father's death, he ascended to the throne and quickly conquered the other Greek city-states and extended his control to the Black Sea, and he soon began to unify the Greeks against their great rivals, the Persians. His initial goal was to avenge an earlier invasion by the Persians against Greece and to restore himself to the Persian throne. He attacked and defeated Darius at Issus and then successfully laid siege to the entire eastern Mediterranean region. He returned to Persia and fought a decisive battle at Guagemala, in which the larger Persian army was defeated and the capture of the cities of Babylon and Persepolis ensued.

Had Alexander stopped at this point, his legacy would not have been as large in history. Alexander had, even before his conquest of Darius's dominions, been offered the Western half of the Persian empire, which his friend and general Parmenion is said to have argued "I would accept the proposal if I were Alexander," and to which Alexander famously replied "So would I, if I were Parmenion." Nor was Alexander content to be considered "king of the Persians;" rather, he wanted to become "king of Asia" and ultimately "king of the World." He aimed to conquer the entire extent of the world, which Aristotle believed lay as far east as the Indus River. Alexander's armies traveled as far north as the Samarkand region of central Asia, crossed the Hindu Kush in a remarkable campaign and then delved south to cross the Indus River in what is now Pakistan. Alexander's men eventually rebelled against going farther, and Alexander ultimately knew that he had not conquered the entire world east of Greece. But, he had built an empire greater than anyone before him and had brought about an enormous exchange of trade and culture as the Asian and European cultures made their first significant encounter.

---

[8] Most of this account comes from *Alexander the Great: The Invisible Enemy* by O'Brien (1992), *The Genius of Alexander the Great* by Hammond (1997), and *Alexander the Great*, a web biography by Jona Lendering (www.livius.org/aj-al/alexander/alexander00.html).

Alexander's greatness resulted not only from the size of his conquests or the daring and brilliance of his military strategies – reflected in the fact that his victorious army was often outnumbered by native armies in the Asian campaigns – or his ability to form alliances. Alexander possessed a greater vision than any of his contemporaries, as he undertook a policy of fusion, both economically (through trade) and socially (through intermarriage). He was a great proponent of Greek culture and he helped to transmit it far and wide throughout Asia, and he trained Asian soldiers at a young age in Greek culture and martial arts. But, he also adopted many Asian cultural elements, let Asian satraps run much of the empire, and even wanted to make Babylon the future home of his empire. His brilliance, strategic vision, and tremendous sense of personal destiny must be regarded as some of his more "positive" dopaminergic traits.

Unfortunately, Alexander's legacy was tarnished by the consequences of his more negative dopaminergic traits. Whatever more noble aims may have inspired his original conquests were more than offset by ruthlessness and, in later years, a paranoia fueled by alcohol. One example of this was the beachfront crucifixion of 2,000 men of Tyre (a city in southern Lebanon), which had resisted his army for several months before falling. Other examples were his wholesale destruction of Persepolis, which he later somewhat regretted, the wholesale slaughter of Greek expatriates in India for a cultural transgression, and his senseless purges of administrators following the return to Persia after his disastrous return trip across the Makran Desert, in what is now southern Iran. His daring was accompanied by a restlessness and even stubbornness, which led to the disastrous Makran voyage, in which by some accounts almost 80 percent of his army may have perished. Finally, his tremendous belief in his abilities and destiny was offset by a mystical, megalomaniacal streak in which he cultivated an almost god-like image of himself. As a member of the Macedonian royal family, he could somewhat "legitimately" claim to be descended from demigods such as Hercules and he acquired royal lineage during conquests (such as being the heir to Ammon in Egypt), but these were mythical and titular links. Alexander's mysticism, however, took the form of a much more grandiose self-deification, which led to ridicule and even a serious near-mutiny by his Macedonian troops in one instance, which he quelled by supposedly accepting his mortal lineage in front of them in one of his more masterful scenes.

Alexander's early death at the age of thirty-two led to the rapid division of his physical empire, but the cultural unifications brought about by his conquests would continue centuries after his death and his legacy would especially influence the psyche of ancient Rome.

6.2.2   Christopher Columbus

Christopher Columbus ("Colon" in Spanish) (1451–1506) certainly ranks as one of the most influential and mysterious persons in history.[9] Despite his so-called "discovery" of the Western Hemisphere – arguably the single-greatest event in all of history – Columbus's life is riddled with enigma, including his birth date (probably late summer or early fall of 1451), birthplace (widely but not universally accepted as Genoa, Italy), and his religion (alternatively, Jewish by family with later conversion to Christianity, or pious Christianity throughout). However, as described by Wilford (1991), the driving force behind Columbus's obsession with finding a Western passage to the Far East has befuddled historians most of all.

What is known is that Columbus received throughout the course of his life a more than adequate education for his times. He left in his early teens for a life at sea, initially as a Portuguese sailor, traveling widely along the coastlines of Europe and Africa, including the Canary Islands, the westernmost extent of the Spanish empire. Columbus gained a good knowledge of geometry, astronomy, and cartography, as would have been required of navigators, and he became quite proficient at several languages, including Latin. It is not clear when or how Columbus first entertained the belief that a westward course to Asia could be achieved. There were previous indications of a populated land mass to the west, because westerly winds and currents (so named because of their origination in the west) would occasionally carry artifacts and even occasional dead bodies to the islands lying off the western coast of Africa (i.e. the Azores, Canaries, and Madeiras). There was also the belief of the Italian scholar Toscanelli, with whom Columbus corresponded, that a westward passage could succeed, and there were even apocryphal stories of an encounter with a mariner in the Madeira Islands who secretly told Columbus of being blown off-course in a westward direction and briefly landing on a large land mass. But, most of these clues would have been available to other sailors, yet only Columbus developed the single-minded obsession to sail westward. Hence, researchers have concluded that only by understanding Columbus' personality can his great quest and discovery be understood.

Columbus knew that, as the Portuguese methodically extended their voyages along the coast of Africa, a maritime route to the Indies and their fabulous wealth would eventually be discovered. But, he was

---

[9] Much of the following synopsis is based on Morison's *Admiral of the Ocean Sea: A Life of Christopher Columbus* (1983) and Wilford's *The Mysterious History of Columbus* (1991).

restless and could not wait for such an eventuality, which because of its length might in any case prove unprofitable relative to the overland routes already in existence. Columbus recognized the great consequences of sailing westward to the Indies and totally convinced himself, for what in retrospect turned out to be conveniently flawed reasons, that it was feasible to make such a voyage.[10] Partly there was the allure of riches, but there was also the delusion that he was chosen by God to be the instrument by which the westward route would be discovered, and that his exploits would lead to great wealth and power that Spain would eventually use to reclaim the Holy Land. Columbus relayed these delusional and even paranoid thoughts in the *Book of Prophecies*, drafted in the 1499–1500 timeframe before his fourth voyage, a work that many historians dismissed as the result of Columbus' poor mental state following the humiliations suffered after his initial voyages. But, West (cited in Wilford, 1991) as well as Wilford himself argue convincingly that the seeds of his delusions and mysticism were present long before his initial voyage to America and reflected a core element of Columbus's personality – a delusion of grandeur combined with an intense mysticism, going far beyond normal piety (Wilford, 1991).

This delusional and mystical side of Columbus coalesced with his other great dopaminergic traits – intelligence, restlessness, and obsessiveness. As regards his intelligence, Columbus had a keen mind, a variety of intellectual interests, and a successful self-education. Columbus was prone to making creative although often outlandish and even comical hypotheses and associations, and he was a brilliant if somewhat intuitive navigator.[11] His restlessness is well-documented, both before and after his initial voyage, but his dogged pursuit of the dream of sailing westward is a testament to both the strength of his delusions as well as his inordinate goal-directedness.

Like most great historical obsessions, Columbus's vision was conceived in his youth. His first formal petition to mount a westward

---

[10] Columbus assumed that East Asia extended further eastward than actually was the case, that the Earth was 25 percent smaller than in reality (because he accepted the low-end of contemporary estimates of the conversion of degrees to nautical miles), and there were likely to be as-yet undiscovered islands that would serve as stepping stones to the East Indies. His final estimate of the distance from Spain to Japan (then known as Cipangu) was only 2,400 nautical miles, or about 25 percent of the actual distance!

[11] Morison (1983) proposed that, because of their enormous expertise at both ocean and inland sailing, Columbus and James Cook rank as the greatest navigators of all time. Columbus' eight remarkably successful transatlantic voyages and his explorations of uncharted lands and waters in the space of a decade arguably rank as the greatest maritime feat in history. Morison provides example after example of Columbus's genius for dead-reckoning and his amazing ability to negotiate and overcome a variety of formidable obstacles on his various sails.

expedition was made to his native countrymen (the Genoans), when Columbus had just turned thirty. He then unsuccessfully petitioned the king of Portugal for a fleet to sail westward in 1483, at the age of thirty-two. For the next decade, his determination never failed, and in 1492 he finally was successful in obtaining his first expeditionary fleet from the Spanish throne, along with a claim to enormous potential riches and titles. Columbus experienced extraordinary luck on his voyage, unknowingly hitting the easterly tradewinds on the outbound journey and the westerlies on the return sail. He and his crews made landfall after passing the 3,400 nautical mile point (roughly 1,000 nautical miles beyond his original estimate of the distance of Japan), just after he had agreed to turn back if no landfall was sighted by the very next day. Columbus eventually reached the Western Hemisphere in October of 1492 in the Bahamas chain, probably on what was formerly known as Watling Island and since rechristened San Salvador Island. Columbus would make four voyages in all, and he was the first modern European to discover Central and South America as well as every major island in the Caribbean. Although Columbus failed in his effort to find a passage to the Indies, even he eventually began to realize he had discovered a new continent in the process.[12]

Columbus' discovery had an astounding psychological impact on Europe and is said to have inspired as much as any other event the transition from medieval Europe to a new age of exploration (Morison, 1983), which in turn inspired a great many scientific and mathematical discoveries. Columbus achieved an even greater iconic status in the emerging United States, with a host of American cities, rivers, universities and other landmarks named after him. For centuries, Columbus's discovery was regarded mostly in a triumphal light, but the darker side of his dopaminergic personality led to consequences that are now regarded as tragic and even catastrophic. While his first encounter with the Native Americans of the Caribbean was relatively benign, his second one led to the formation of a colony (Isabel, on the island of Hispaniola) and enslavement of those natives who resisted the Spanish. Columbus does

[12] Much is made of Columbus's supposed "blunder" in not realizing that he was nowhere near the East Indies. Columbus did recognize that he had discovered a new continent in South America, but even after his final voyage there was no formal proof (although there were lots of suggestions) that he was not in the vicinity of the East Indies. My general opinion of Columbus echoes that of leading historians in that he had a dual personality – a medial dopaminergic one that was propelled by mystical and cosmo-graphical ideas and an almost unshakable belief that he had reached the East Indies along with a lateral frontal dopaminergic system that supported his impressive empirical observations and reasoning and would have eventually led him to the realization of his correct geographical location.

not bear full responsibility for the tragedy that unfolded, being himself initially fair in his treatment of the natives, who he merely wanted to convert, not punish. However, once conflict with the natives had been initiated by others, Columbus became extremely harsh in his treatment of the natives and even somewhat harsh in commanding his own people, the latter treatment leading to a series of mutinies and his subsequent imprisonment at the end of his third voyage. The harsh tributes he demanded of the natives and their subsequent enslavement, partly to provide some measure of commercial return,[13] was particularly egregious, and most of the slaves he brought back died on their way to Spain. In the end, it was estimated that of the original 300,000 natives living on the island of Hispaniola (now containing the Dominican Republic and Haiti), a third perished during the first few years of Columbus's governances and only about 15,000 survived twenty years after the initial Spanish conquest (Morison, 1983). This was a prelude to an even more ghastly fate awaiting the natives of the American mainland, whose population after a century of Spanish rule was almost totally decimated, from a pre-conquest estimate of fifty million to but a few million in the end.

6.2.3   Isaac Newton

If Columbus's discoveries sounded the beginning of the end of the Medieval period, the discoveries of Isaac Newton (1642–1727) represented a crushing, final blow. His brilliant discoveries and analysis led to the first-ever systematic understanding of the cosmos and two of its most mysterious elements – light and gravity – and cemented his rank as arguably the most influential scientist and mathemetician in history.[14]

Newton showed brilliance as a student early on, and it was in his late teens that it became clear that he was better suited for academia than for the farm (which nonetheless provided him some practical training early on). Newton was admitted to Cambridge University at the age of eighteen and quickly mastered the principles of Euclidian geometry, but it was shortly after graduation at the age of twenty-two that he conceived his great ideas about light and gravity and even the calculus (which he

---

[13] Columbus was aware that, because of his failure to find major gold deposits and other commercially valuable goods in Hispaniola, not only his future wealth but also his future explorations funded by the Spanish monarchs were in jeopardy. This was a major factor that led to his resorting to slave-trading as an alternative source of revenue.

[14] This account of Newton is distilled from numerous sources, including *Isaac Newton* by Gleick (2003), *Newton's Gift* by Berlinski (2000), *Isaac Newton Reluctant Genius* by Ipsen (1985) and the more critical source *Newton's Tyranny: The Suppressed Discoveries of Stephen Gray and John Flamsteed* by Clark and Clark (2001).

termed "fluxions") in what has been termed a "miraculous" year. His prodigious intellectual creations occurred largely in isolation, when he had returned to his small town after the plague had once again swept through the cities of Europe. Newton, like many scientists, worked mostly alone, but the isolation throughout his early life was largely self-imposed and so extreme that many of his ideas were published long after their creation and only after much coaxing by friends, or were eventually published by someone else (e.g. the calculus and some of his Bible theories). He was personally somewhat unkempt (though less so when he was working for the British Mint) and he would get so absorbed in his distant and abstract thoughts that he would often fail to eat or sleep.[15] Yet, Newton could also be very practical in that he invented numerous devices, including the sextant and the best telescope of his time (based on reflection), and engaged in a lifelong, hands-on pursuit of alchemy.

His two great works – *The Mathematical Principles of Natural Philosophy* (known by its Latin moniker *"Principia"*) and *Optiks* – represent his lasting achievements, although both were published long after his initial ideas (the first volume of *Principia* was published at the age of forty-four and *Optiks* at the age of sixty-two). The *Principia* in particular arguably represents one of the two greatest scientific triumphs in history – the other being Einstein's general theory of relativity, whose equations were published in 1916. *Principia* offered a unified explanation of a host of terrestrial and astronomical phenomena in the form of the universal law of gravitation, which was a stellar example of dopaminergically mediated creative association of "remote" phenomena. Newton presented his concepts not only in scientific jargon but also in the form of mathematical equations, which had previously been mainly limited to astronomy. It required Newton's combined genius as a rigorous experimentalist as well as the greatest mathematician of his time.[16] Newton, as both scientist and mathematician, did not hesitate to tackle the abstract concept of infinity, which Descartes had shied away from. By unifying physics and mathematics, Newton's fame spread far and wide, and he became the first scientist ever to be knighted by the king of England.

Alchemy, along with his lifelong interest in religion and biblical events, represented Newton's medieval passions. Despite supposedly spending more time on these pursuits than his science and mathematics – he

---

[15] One famous instance of Newton's absent-mindedness is his failure to remount a horse after a steep hill and then leading it by its reins for several miles on foot, without rational explanation.

[16] Newton is believed to have routinely solved geometrical problems that other great mathematicians required months to work out or had never previously been solved (Ball, 1908).

wrote over a million words concerning the Bible, most of which were unpublished in his lifetime – his work in this area was of little use to modern chemists and biblical scholars and historians. He was adamant about debunking the divinity of Jesus – and his obsessive and futile attempt to combine mathematics and biblical ideas led to what seems in retrospect to be a mystical perversion of his great mathematical genius.

Newton's dopaminergic brain allowed him to see farther into the fundamental forces of nature and the cosmos than anyone else in his time and gave him the obsessiveness to pursue these ideas, even to his own bodily detriment at times. But, Newton's dopaminergic brain also negatively affected his personal relations. He was required to be celibate under the terms of his professorship at Cambridge University, but he was much more of a loner than custom required. He was cantankerous, even somewhat antisocial as a youth, and as an adult often squabbled with fellow scientists. For example, his dispute with Leibniz over the invention of the calculus was one of the most infamous scientific quarrels in history, and his disputes with the great scientist Hooke and the astronomer Flamsteed also reflected extremely badly on Newton, who ran the Royal Society of London in a heavy-handed and autocratic manner. He was also inordinately secretive (as witnessed by his cryptic challenge in an early letter to Leibniz), which led him to delay or even fail to publish many of his most important findings. As Warden and later Master of the Mint, Newton did not display sympathy toward counterfeiters, hounding them and even referring many of them to the gallows over the objections of more magnanimous colleagues. Some psychologists such as Baron-Cohen (see Muir, 2003) have suggested that Newton had Asperger's syndrome, but there is no evidence that his social skills were so disturbed as to qualify for this diagnosis; on the contrary, Newton had extensive governmental service (including a couple of stints in Parliament) and a longstanding tenure as president of the Royal Society. Beginning about the age of fifty, Newton had a well-documented psychosis lasting more than a year, which some attributed to mercury poisoning from his alchemy experiments – he was later found to have highly elevated mercury levels in his hair – but his psychotic symptoms are not consistent with those of mercury poisoning (Spivak and Epstein, 2001). Nor are they consistent with schizophrenia, which generally occurs much earlier in males and tends to be more chronic.[17]

---

[17] Much speculation continues to surround Newton's psychosis, which in addition to mercury poisoning has been considered to be brought on by such factors as depression from the alleged breakup of a secret romance or to the aftermath of his *Principia*, published five years earlier.

The clinical traits that most closely describe Newton's behavior are a schizoid personality or perhaps even schizo-affective disorder, which combines introversion with paranoia. Clearly, Newton frequently also manifested obsessive-compulsive and manic symptoms, and he is reputed to have exhibited bipolar symptoms as well (Papolas and Papolas, 2002).

Despite his various disputes and personal failings, Newton in the end also expressed a sincere humility, perhaps more so than the other dopaminergic figures reviewed here. Part of the failure to publish many of his greatest ideas may have stemmed from the fact that public acclaim never interested him as much as it did some of his peers. Newton's humility is exemplified not only in his famous statement that he could only see farther because he stood "on the shoulders of giants" (supposedly referring mainly to Kepler and Galileo) but also in the self-epitaph he offered at the end of his illustrious life:

I seem to have been only like a boy playing on the seashore, and diverting myself in now and then finding a smoother pebble or prettier shell than ordinary, while the great ocean of truth lay all undiscovered before me. (Ipsen, 1985: 86)

6.2.4   Napoleon Bonaparte

Napoleon Bonaparte (1769–1821) represents one of the most remarkable and controversial figures in history.[18] On the positive side, he was a military genius with a prodigious memory and a manic capacity to work (often up to fifteen hours per day), whose triumphs (especially at Austerlitz) rank among the greatest in military history and who, at least early on, represented the prototype of the benevolent despot, whose strength of governance lay in his ideas and leadership as much as his military might. Although the medieval intellectual world had already succumbed to Newton, Napoleonic rule in Europe led to the disintegration of the last vestiges of the medieval social order, in which monarchs and feudal aristocracies reigned supreme. On the negative side, Napoleon was consumed by an overwhelming ambition that led to arguably the greatest single military defeat in human history, and he reputedly suffered not only from temporal-lobe epilepsy (which is associated with various delusions, including delusions of grandeur) but perhaps bipolar disorder as well.

---

[18] This account was based on numerous sources, including Johnson's *Napoleon* (Johnson, 2002), *A Brotherhood of Tyrants* by Hershman et al. (1994), and *Napoleon and the French Revolution* by Holmberg (2002).

Napoleon's character and legacy have many parallels with those of Alexander the Great, in that they were both intelligent, restless and even grandiose risk-takers who conquered most of the known or advanced world of their day by their early thirties, changed forever the societies of their conquered lands, and suffered great military disasters before their flames were extinguished. They even shared a heralded conquest of Egypt and two very shameful acts of ruthlessness within a short distance from each other: Alexander's slaughter of the citizens of Tyre and Napoleon's slaughter of prisoners at Jaffa (Haifa). Although he himself was somewhat pragmatic, Napoleon's more lasting historical influence lay not as a military conqueror but as the force that carried the ideals of the French Revolution (liberty, fraternity, and equality) across the face of Europe. Everywhere he went, the Revolutionary legal code – known for a time as the Napoleonic Code, still the basis of law in most of the advanced nations in the world – overthrew the hereditary rights of the feudal aristocracies, and Napoleon's conquests were accompanied by the establishment of public schools, museums, continental road systems, the decimal system, and even such relatively trivial changes as driving on the right-hand side of the road.[19] Only in Russia, the only major continental country that remained largely unconquered by Napoleon, did feudalism remain until the twentieth century.

However, Napoleonic progress carried huge costs – over one million dead soldiers during the Napoleonic wars. Napoleon cannot be blamed solely or even in the main for these wars, as revolutionary France was constantly under attack or blockade by England and monarchic neighbors such as Austria. But, Napoleon famously exhibited little concern over the loss of life in his military campaigns, once claiming to the diplomat Metternich that the loss of a million men meant nothing if it led to triumph. As Madame de Stael wrote about Napoleon's lack of emotional attachment:

He is a chess-master whose opponents happen to be the rest of humanity . . . Neither pity nor attraction, nor religion nor attachment would ever divert him from his ends. (cited in Johnson, 2002: 115)

---

[19] Because the French nobility had continued in the Roman legion tradition of moving on the left-hand side of the road, which was originally advantageous for protecting the body from the left side because the Roman shield protecting the heart was held with the left hand, the Revolution established that all post-monarchical Frenchmen would travel on the right side of the road. England and Scandinavia, which were never conquered by Napoleon, remained European countries resistant to the rightward travel direction, although Sweden eventually became the last nation in continental Europe to switch rightward in 1957. It turns out that, because of postural reflexes, right-sided vehicular travel may actually be safer for right-handed individuals (Coren, 1992: Chapter 14).

His emotional detachment, along with his delusions of grandeur, led to the greatest military disaster of all time – the Russian campaign, in which his massive army of over 400,000 strong ended up losing over 95 percent of its number, with about half taken prisoner and the other half victims of combat, disease, and starvation. Surprisingly, all was not lost – Napoleon was able to reorganize his army and go on to a string of victories and was even offered generous peace terms by Metternich in which he could have kept many of his original conquests. But, like Alexander before him, what would have seemed generous to a less grandiose person was not acceptable to Napoleon. He eventually lost all, recovered for a brief triumphal return, and then finally lost all again permanently at the Battle of Waterloo in Belgium. By then, however, Napoleon had done too much damage to the traditional social order, and the victorious aristocracies were soon swept aside by the modern nation-states of Europe that eventually came to adopt much of the egalitarianism of the French Revolution, which Napoleon championed but did not always observe in his deeds.

Psychohistorians debate why Napoleon's magnificent abilities were consumed by his dark underside – was it his temporal-lobe epilepsy (which is often accompanied by grandiose delusions, whether of a religious or, as in Napoleon's case, secular nature) or was it bipolar disorder (Goodwin and Jamison, 1990; Hershman *et al.*, 1994), leading him to great accomplishments during his manic periods and to irritation, negative outbursts, and even ruthlessness during his depressive outbursts? In the end, the answer may not be especially relevant in that both disorders are associated with excessive dopamine.

6.2.5   Albert Einstein

The physicist Albert Einstein (1879–1955) was the pre-eminent scientist of the twentieth century and arguably the greatest scientist since Newton, at least in the eyes of the newsweekly magazine *Time*, which deemed him the most influential person of the twentieth century.[20] Although Einstein's thoughts helped to overturn the classical physics originated by Newton, his and Newton's lives were close to being historical twins (Lightman, 2004). Both men had a love of physics (Einstein

---

[20] (www.time.com/time/time100/poc/magazine/albert_einstein5a.html). Much of the material in this section was distilled from *Einstein: A Life* by Brian (1996), *Einstein: The Passions of a Scientist* by Parke (2003), *Einstein: His Life and Universe* by Isaacson (2007), and the single-mindedly scathing account *Albert Einstein: The Incorrigible Plagiarist* by Bjerknes (2002).

prided himself on his great intuition whereas Newton was a more practical inventor and a much more gifted mathematician),[21] both focused on light and gravity as their two major scientific interests in life, both had a "miraculous" year in their early-to-mid twenties in which they developed at least three profound ideas (Newton in 1666; Einstein in 1905), both were somewhat withdrawn and emotionally distant from friends and family (Newton much more so than Einstein), both were absent-minded and somewhat disheveled in their appearance, both periodically engaged in manic, obsessive periods but suffered serious physical and mental collapses in middle age, and both frittered away much of their later life pursuing scientific goals (Newton his alchemy and biblical history, Einstein his unified field theory) that were either not possible to achieve in their times or perhaps any time.

In Einstein's miraculous year of 1905, he managed to publish four scientific papers that established his reputation as one of the leading physicists in the world and that eventually led to his award of the Nobel Prize in 1921. These related to the photoelectric effect(law), Brownian motion, the measurement of the size of molecules, and his famous paper on special relativity describing the effects of relative motion, the unique nature of light and how it, and not time, were constant in the universe. In much of his earlier work, Einstein succeeded in unifying previously separate approaches (wave analysis vs. particle/statistical analysis) to account for important phenomena (Isaacson, 2007), which again represents the power of remote associations in dopaminergically mediated creative thought. Although Einstein had been thinking about special relativity for years and extensively borrowed (unfortunately, without acknowledgement) from previous work of Poincaré, Lorentz, and possibly even from the Italian physicist De Pretto, who actually first derived the famous formula $E = mc^2$ in 1903 (Bjerknes, 2002), it was Einstein who arguably most fully grasped the metaphysical implications of denying the absolute status to space and time accorded them in the Newtonian universe. Einstein quickly moved on to tackle the problem of gravity in relativistic terms, and over the ensuing years leading up to the formal publication of his general theory of relativity in 1916, Einstein (with crucial help from his old Zurich colleagues Minkowski and Grossman and the mathematician David Hilbert) contributed to the equations of the four-dimensional space–time continuum. Einstein's

---

[21] Indeed, Einstein was supposedly described by the Nobel prize-winning physicist Eugene Wigner as a "terrible mathematician," although that characterization was presumably relative to physicist standards.

general relativity theory is widely held alongside Newton's *Principia* as one of the two greatest achievements of the human mind.[22]

The enormous effort placed into his work during this period was not beneficial to his marriage to fellow physicist Mileva Maric; but, even in their earlier romantic interlude, Einstein's letters to her dealt more with his scientific ideas than their relationship. Einstein's preference for work over personal relationships manifested itself in a second strained marriage (to his cousin Else) and an estranged relationship with his two sons. Einstein's physical restlessness (begun during the numerous family and schooling moves in his youth) did not help, as he moved from Zurich (where he had settled into an extremely productive academic life) to Prague and then Berlin, both of which made Mileva depressed and which arguably didn't improve his scientific productivity until his fateful encounter with Hilbert in 1915. He enjoyed the company of women, but his many affairs appear to have been casual ones, with little emotional attachment. Nevertheless, although his lack of emotionality impaired his ability to become deeply involved with specific humans, Einstein loved "humanity" in the abstract and evidenced much more political idealism than Newton by campaigning for pacifism and socialism and a Jewish state in his middle and later years.[23]

While Einstein's intellectual strengths and personal weaknesses clearly are hallmarks of a dopaminergic personality, the question of whether Einstein actually exhibited any hyperdopaminergic pathology is not clear. It has been argued that Einstein, like Newton, may have been mildly autistic because he was language delayed and had poor social communication skills (Muir, 2003). But, actual evidence from family accounts indicate that Einstein was not language delayed, nor was he dyslexic, and he was always an above-average to even outstanding student, at least in the subjects he liked. Despite a certain stubbornness and even arrogance in his youth, Einstein's social skills were adequate and he had many friends and acquaintances throughout life. He unquestionably suffered from manic and obsessive periods in his life where he would be consumed by his scientific pursuits and sleep and eat poorly, and he suffered several instances of at least mild depression: for example, when he was left alone at age fourteen in Switzerland to attend school; when

---

[22] Although it is clear that the great German mathematician Hilbert influenced Einstein in formulating the equations of general relativity, which are widely known as the Einstein–Hilbert equations, even Hilbert always acknowledged that the basic theory itself was a tribute to Einstein's intuitive genius (Todorov, 2005).

[23] Einstein once remarked that "My passionate sense of social justice and social responsibility has always contrasted with my pronounced lack of need for direct contact with other human beings" (Lightman, 2004: 108).

he couldn't find a job after graduation from college; after he had completed his sensational papers in 1905; and after having published the general relativity theory in 1916, during which time he suffered from a serious eating disorder that led to the loss of almost sixty pounds and confined him to long bed-rests. Although such occurrences could in some respects suggest that Einstein suffered from bipolar disorder, his workaholism and even obsessiveness to the neglect of his personal life suggest a leaning toward obsessive-compulsive spectrum disorder.

Although he was in some ways more scientifically magnanimous than Newton – certainly, he did not initiate any serious scientific feuding – Einstein had perhaps a fundamentally more grandiose belief in his own abilities than did Newton. Although he was less ostensibly pre-occupied by religious matters than Newton, he was a believer in "mystic emotion" and was more convinced than Newton that he could "know God's thoughts." Einstein failed to accept the increasing ascendancy of quantum mechanics because he did not believe that the universe could be probabilistic at heart, leading to his famous statement that "God does not play dice with the universe," which led the quantum physicist Bohr to scold Einstein to "quit telling God what to do." And, ultimately, whereas Newton could dabble in a host of different scientific projects (even managing to revise the *Principia*) long after his greatest triumph, Einstein kept pressing fruitlessly on in attempting an even more gigantic creation (the unified field theory), which neither he nor any other physicist armed with the scientific facts of his day was qualified to solve.

6.2.6   Dopaminergic personalities in history – reprise

Despite their different fields of endeavor, the above giants of history shared many dopaminergic intellectual and personality features in common, as summarized in Table 6.1. All of these men were highly to extremely intelligent and all had far-reaching visions, including a pre-occupation with religious, mystical, or other grandiose themes. These men all were continually restless and had intense motivational drives – a trait that characterizes most of the great achievers in history (Simonton, 1994). Finally, all of these men were rather unemotional, none let personal relationships stand in the way of his larger goals (although Einstein at least escapes the charge of ruthlessness), and all possessed an overconfidence and even recklessness that led them in many cases into major blunders. Whether any of them actually exhibited any clinical hyperdopaminergic disorders is less clear: Napoleon comes closest with his reputed seizures and possible bipolar disorder, although even in this instance the issue seems unclear. It is more likely that all of these men to

Table 6.1 *Dopaminergic traits in famous men of history.*

Person	Positive DA traits	Negative DA traits	Closest DA clinical disorder
Alexander	High intelligence, visionary ideas, high motivation, risk-taking, extreme self-confidence.	Grandiosity, ruthlessness, restlessness, paranoia.	Schizotypy ADHD?
Columbus	High intelligence, visionary ideas, high motivation, risk-taking, extreme self-confidence.	Grandiosity, occasional ruthlessness, restlessness.	Schizotypy ADHD? Bipolar?
Newton	Extremely high intelligence, visionary ideas, high motivation, self-confidence.	Obsessiveness, lack of empathy and social skills, occasional ruthlessness, paranoia, personal neglect.	OCD Bipolar Asperger's?
Napoleon	High intelligence, visionary ideas, high motivation, risk-taking, extreme self-confidence.	Delusions of grandeur, ruthlessness, restlessness.	Temporal-lobe epilepsy Bipolar Schizotypy?
Einstein	Extremely high intelligence, visionary ideas, high motivation, self-confidence (risk-taking?)	Obsessiveness, lack of empathy and social skills, personal neglect, grandiosity.	OCD Bipolar Asperger's?

varying degrees suffered aberrant personalities, ranging from the schizotypal (more characteristic of Alexander, Columbus, and Newton because of their more suspicious natures and, in Newton's case, his psychotic breakdown later in life) to the obsessive-compulsive (most characteristic of Einstein and Newton) and even impulsive (Alexander and possibly Napoleon).[24] The difficulty in assigning a particular disorder to each individual stems partly from the enormous overlap of the various hyperdopaminergic symptoms, as described in Chapter 4. Nevertheless, just as obsessive-compulsive disorder appears to most typify the hyperdopaminergic clinical disorders, so, too, did all the famous individuals reviewed manifest great obsessions and, at times, even compulsive tendencies – Alexander in his attempt to conquer the known world, Columbus in his quest to find a westerly route to Asia,

[24] Alias (2000) argues that great leaders in history have not been schizotypal, but the fact that so many of them have had a great sense of personal destiny and even messianic purpose as well as a healthy dose of suspiciousness indicates otherwise.

Newton in his effort to mathematize the laws of science and to understand the hidden codes of the Bible, Napoleon in his attempt to conquer all of Europe and reign in France's great maritime nemesis England, and Einstein in his quest to find the unified field theory. All of these men combined a large amount of lateral dopaminergic activity – which provided them their scientific inquisitiveness and/or strategic focus – along with the creativity and visionary (even irrational) zeal indicative of a strong medial-dopaminergic activation What is important is not just that these and other leading figures of history had brains that were personally rich in dopamine, but rather that their accomplishments helped create the goal-directed, competitive, restless, somewhat alienated, highly technological, and ever-changing character of our modern, highly dopaminergic societies. For example, there is little doubt that Alexander's conquests increased trade and cultural exchange across widespread areas of Europe and Asia, Columbus's achievements led to a marked increase in exploration and discovery, Newton's intellectual triumphs helped usher in a new wave of science and invention, Napoleon's conquests paved the way for the modern nation-states with their centralized, bureaucratic power, and Einstein's discoveries helped stimulate greater exploration of the cosmos and advanced technologies based on his revolutionary physics, such as nuclear power and lasers. Few would argue that these changes – which managed along with hundreds of others to transform the simpler lifestyles of the hunter-gatherers and agrarians into the fast-paced, competitive, highly technological, and future-oriented hyperdopaminergic societies of today – were not without major benefits to humankind. But, there were costs as well, to both individual humans (including the five highlighted here) and society, as will be discussed briefly in the next section and much more extensively in Chapter 7.

6.3   The modern hyperdopaminergic society

Although they may have manifested higher dopamine activity than the hunter-gatherers, the agrarian societies (including those of today) offer a lifestyle that differs considerably from that of the modern industrialized ones. Despite a certain degree of social stratification and frequent violent confrontations over land and resources, the early agrarian civilizations for the most part were characterized by social, economic, familial, and geographical stability, as well as a deliberate pace of life. In contrast, present-day industrial and post-industrial societies, as well-described by Clark (1989), Toffler (1970), and others, feature long and stressful workdays and commutes, incessant time demands, rapid change, reduced sleep (now estimated at less than seven hours per day), transient emotional/social relationships, compulsiveness in seeking secondary rewards (material goods, status symbols, fame, power, information etc.), intense competition, and financial uncertainty. In turn, the drive to acquire incentive-based rewards and the constant decision-making and control strategies associated with succeeding in modern societies all lead to an extremely high "future-orientation" (Bentley, 1983; Zimbardo, 2002). In contrast to the 3,000 calories and 15,000 calories per day expended per capita by the hunter-gatherers and agrarian societies, respectively, modern societies expend ten calories to produce one calorie of consumed food and have an overall per capita energy consumption of over 300,000 calories per day (Clark, 1989).

What does it mean for a modern society to be labeled "hyperdopaminergic"? Based on the dopaminergic personality profile in Chapter 3, the hyperdopaminergic society is above all, *an extremely goal-driven society* in which achievement is highly rewarded in a highly competitive and uncertain environment. Those relatively few individuals with a high internal locus-of-control, future-orientation and strong motivational drive – until recently, considered masculine traits – are most likely to thrive in such a society.[25] The hyperdopaminergic society is less concerned with the pleasures of consumption per se – e.g. the sensual experiences of eating, touching, or other relaxed endeavors, which are more dependent on serotonergic and noradrenergic systems – because dopamine is not involved in consummatory behavior per se. Rather, the hyperdopaminergic society is incentive-based in its quest for secondary rewards and driven more by abstract ideas such as religious and political domination and scientific truth, all springing from the more primitive dopaminergic drive toward distant space. A highly dopaminergic society is fast-paced and even manic, given that dopamine is known to increase activity levels, speed up our internal clocks and create a preference for novel over unchanging environments. The important contribution of dopamine to abstract intelligence, in combination with the basic dopaminergic involvement in exploration and novelty, leads such societies to be highly technological in nature. Finally, the inhibition of socio-emotional activity by dopamine – as evidenced by the lack of empathy in the highly dopaminergic individuals reviewed in the preceding section and in the poor socio-emotional skills of

---

[25] Taylor correctly points out in his Chapter 9 that the masculine "psyche" cannot be ascribed merely to the male hormone testosterone, because many females in industrialized societies now exhibit these traits, despite having almost no testosterone in their bodies. Rather, masculine psychological traits should more precisely be considered *dopaminergic* and, because testosterone is but one of many influences on dopamine levels in the brain, many women may possess high levels of dopamine and thereby exhibit what have traditionally been considered masculine intellectual and personality traits.

the dopamine-rich left hemisphere – means that dopaminergic societies are typified more by conquest, competition, and aggression than by nurturance and communality.

It is fairly obvious that, based on the above description, technologically advanced, industrialized societies are in most respects much more dopaminergic than are primitive, agrarian societies. As documented by Pani (2000), Previc (2007), and others, the collective tendency of most of the pressures in modern industrial and post-industrial societies – including active coping with stress and uncertainty, constant decision-making, anxiety concerning the future, the presence of constantly changing (novel) environments, a reduction in emotional support, and sleep deprivation – is to increase dopamine in the brain. Not surprisingly, hyperdopaminergic social-deficiency disorders such as autism and schizophrenia are much rarer or at least less severely manifested in non-industrial societies – e.g. autism is almost unheard of outside of the industrialized nations and even in traditional groups within modern societies, such as the Amish in the United States (Olmstead, 2005; Previc, 2007; Sanua, 1983), and chronic schizophrenia is primarily found in the industrialized world (Sartorius *et al.*, 1986). Indeed, it is highly educated and successful individuals in the industrialized world, with presumably the highest levels of dopamine, who are up to four times as likely to produce offspring with autism (Previc, 2007).

As anyone who has transitioned from a nonindustrialized to an industrialized society can attest, one of the most obvious differences is the faster pace of life, constant change and reduced communal orientation in the latter. Indeed, as described in Toffler's *Future Shock*, the constant movement of individuals across residences, jobs, and the like makes it difficult to maintain normal social relationships in modern societies. For those individuals unaccustomed to the faster pace of life, the adjustment from a traditional society to a modern one can be quite painful, with inadequate dopaminergic coping skills and reduced social support frequently combining to produce depression ("learned helplessness," according to Seligman, 1975). Even after several centuries of contact with modernity, many indigenous peoples are caught between their ancestral traditions and the demands of modern societies, often with disastrous consequences. However, it can be equally difficult for members of industrialized societies, particularly if they are younger and at their dopaminergic peaks, to suddenly transition to a more primitive society or slower pace of life. Such a transition, exemplified by the well-known "island fever" following confinement to a laid-back tropical island lifestyle, is akin to the withdrawal from other addictive behaviors that dopamine mediates and can be associated with extreme boredom and anxiety (Taylor, 2005).

The seeds of the modern hyperdopaminergic society were planted early in the European Renaissance by Newton and Columbus and a host of other great scientists and explorers. They were nurtured by the rise of new individualistic and achievement-oriented value systems (e.g. the so-called Protestant work ethic), the highly competitive capitalistic societies that emerged in concert with the industrial revolution, and eventually by the rise of the nation-state. But, it wasn't until the mid-point of the twentieth century that dopaminergic values predominated in the mass culture, in women as well as men, and hyperdopaminergic disorders began to appear throughout the industrialized world. At the beginning of the last century, over 60 percent of the United States population lived on farms, less than 20 percent of the labor force was comprised of females and, despite many advances in health and sanitation, the average lifespan in the most developed nations was in the mid-to-upper thirties – only marginally longer than in Roman times (Pearson Education, Inc, 2009;[26] U.S. Census Bureau, 2004). By the end of the twentieth century, however, nearly 80 percent of the United States' population lived in urban areas, the percentage of women in the workforce increased to almost 50 percent, and the average lifespan in almost all industrialized nations exceeded seventy years.

Even today, industrialized societies are not equivalent in the extent of their dopaminergic drives. By far and away, the United States is the epitome of the hyperdopaminergic society and has been for many decades, being characterized by a particularly fast-paced, highly urbanized, and highly competitive and uncertain environment.[27] The United States is (or certainly has been until recently) the most technologically advanced as well as the most future-oriented of all societies (Zimbardo, 2002). The United States requires individual initiative perhaps to a greater extent than any other nation (e.g. workweeks are long, vacation time is typically limited to one to two weeks, there is no universal health coverage, and there is only limited, unpaid time off for pregnancies), but it arguably rewards individual initiative and success to a greater extent as well. For example, the relative tax burden on the richest of Americans is far less than in other industrialized nations, and the distribution of wealth is less equal than in any other advanced nation. Because of the high divorce rate, extreme time pressures, and transient lifestyle (the average American changes residency at least a dozen times in his or her lifetime), social support is reduced relative to other industrialized

---

[26] www.infoplease.com/ipa/A0104673.html.

[27] As but one example of this uncertainty, over 20 percent of Americans have no health insurance and are but one major medical illness away from personal bankruptcy.

nations and certainly relative to more communal hunter-gatherer and agrarian societies (Toffler, 1970). The United States is also by far the nation that has militarily, religiously, and scientifically thrust itself beyond its own borders, with troops in over ninety nations, missionaries around the world, and scientific conquests extending from the outer reaches of the solar system to the innermost workings of the atom and body (including the recently completed human genome project). The conquest mentality is also apparent in the relationship of the United States to the environment, as it holds less than 5 percent of the world's population but consumes over 25 percent of its resources. Much of the economic activity carried out in the United States and, to a lesser extent, other industrialized countries, is of a "linear" variety in which resources are exploited, transported, refined, consumed, and discarded in an economically aggressive but environmentally inefficient and unsustainable fashion (Capra, 2003).[28] The next closest modern society to the United States in the above respects is Japan, but even Japan remains a more structured, conservative, and less mobile society with a much stronger social welfare system.[29] Unfortunately, the United States has also become the nation with the highest rate of dopamine-related disorders, in particular autism.

6.4   Summary

The progression from the first hominids to the hunter-gatherer and early agrarian societies and ultimately to the modern industrialized societies was propelled by the increasing predominance of dopaminergic systems in the human brain and by individuals exhibiting especially high dopaminergic activity. As depicted in Figure 6.2, the dopaminergic progression was punctuated by four major events:

1. a dramatic increase in thermal tolerance and endurance capability and the introduction of meat into the diet with the advent of *Homo habilis*;
2. the cultural "Big Bang" around 80,000 years ago that roughly coincided with the first widespread consumption of shellfish;

---

[28] According to Capra (2003), a more sustainable "circular" (and arguably less dopaminergic) environmental relationship emphasizes local production and recycling of resources (see Chapter 7).

[29] Besides Japan, the major "dopaminergic" rivals in the last century were Germany and the Soviet Union, both of which were technologically advanced and in possession of large, ruthlessly developed empires built on military might and political ideals (nationalism/fascism and communism, respectively).

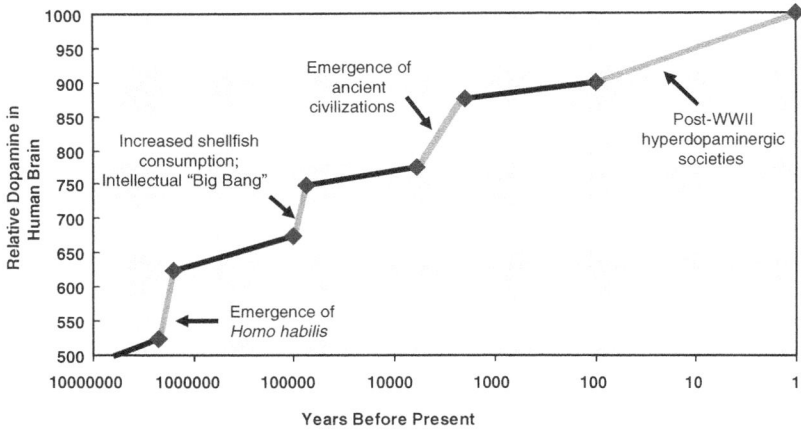

Figure 6.2 The progression of the dopaminergic mind.

3. the emergence of stratified, patriarchal civilizations beginning around 6,000 years ago; and
4. the widespread dissemination of dopaminergic values in the twentieth century, leading to the advent of the hyperdopaminergic societies after the end of World War II.

While all of these stages resulted in significant cognitive advances, only the first of these was part of a significant physiological adaptation that probably included both genetic and nongenetic elements, and only the first two produced clear and relatively equal benefits across the entire tribe or society. The final two dopaminergic stages are documented in the historical record and led to increased competitiveness, dominance of the "left-hemispheric" style, exploration to the farthest realms of the Earth and beyond, exploitation of the resources of the Earth, explosions of artistic creation, and scientific discoveries that uncovered much of the mystery of the universe and of life itself. But, the final dopaminergic epochs also came with an enormous cost, in a series of endless wars and conquests, alienation of humans from the land, stratification of society, inequality between the sexes, a reduction of leisure time, increased detachment and even alienation from our fellow humans, increased mental illness, and more recently, serious environmental degradation (Taylor, 2005).

The issue to be considered in the final chapter is whether the dopaminergic imperative that has so dominated human evolution and especially its more recent history must now be regarded as counterproductive and in need of relinquishment.

7 Relinquishing the dopaminergic imperative

It would be a dramatic understatement to say that we are at a turning point in human history. Not only for the first time in our history are humans being forced to use increasingly expensive energy resources and raw materials from depleted terrestrial and aquatic resources (Duncan, 2005), but for the first time leading thinkers are beginning en masse to argue for a reconsideration of the values that have dominated human existence since the emergence of the ancient civilizations. In this final chapter, I will highlight the downside of our dopaminergically dominated past history and describe the changes that must occur in industrialized societies in particular to restore balance to our individual and societal existences. Many of the themes in this chapter echo those expressed by other leading advocates of social and environmental change such as Capra (2003), Clark (1989), Duncan (2005), Primavesi (1991), and Taylor (2005).

Even if we chose to, it is almost inconceivable that we could entirely discard our dopaminergic minds altogether, as it would mean reversing over two million years of human evolution. Although the cultural achievements of the first genetically modern humans 200,000 years ago do not appear dramatically different from *Homo erectus*, as is consistent with the proposed epigenetically mediated rise of dopamine over the past 100,000 years, the dopaminergic gulf between *Homo habilis* and its predecessors included a major physiological adaptation that presumably involved some genetic selection and would accordingly be difficult to fully recreate. What can occur, however, is the relinquishing of the *dopaminergic imperative* – the unquestioned sanctity of the human drive to explore, discover, acquire, and conquer/control.

7.1   Reaching the limits of the dopaminergic mind

Very little introduction is needed to describe the accomplishments of the human race. Among the many items on the plus side of the human ledger, we have:

1. achieved enormous feats of exploration on and under our planet and into the immediate space beyond;

2. developed great worldwide religions that, in theory at least, offer moral solutions that promote co-existence among humans;
3. transformed the raw environment into an enormous array of chemicals that sustain our modern world;
4. created drugs and other medical advances that prevent most serious diseases and have dramatically extended our lifespan;
5. built dams that harness mighty rivers and help irrigate the world's deserts;
6. designed and built vehicles that quickly traverse the globe with goods and services and provide a high degree of personal mobility for hundred of millions;
7. made scientific discoveries spanning the cosmic and atomic levels and striking at the mystery of life itself; and
8. created an extraordinary output of literature, music, and art.

Yet, for all but one of the above achievements as well as a host of others, there is a corresponding past, present-day and/or pending "Cassandra"-like consequence (Clark, 1989). For example:

1. exploration brought conquest, slavery, colonialism, the spread of disease and destruction that, among things, reduced the native populations of the Americas by over 90 percent following the European invasion and led to the extinction of more than half of the animal and plant species in the world and is now threatening even the upper atmosphere with tens of thousands of pieces of "space junk";
2. religious and political ideologies fostered differences that led to great wars and, in some cases, rationalized and even encouraged the eradication of other humans, species and much of the natural environment;
3. the enormous increase in synthetic materials has led to a massive increase in toxic pollutants in our homes and streams and air and to large increases in various cancers and chemical sensitivity disorders such as asthma and allergies (Ashford and Miller, 1991);
4. the miracle drugs that once controlled disease have brought about immune resistance that now makes them largely obsolete,[1] and the artificially induced longer lifespans have produced overpopulation and other problems that are, turn, leading to the reverse scenario of shortened lifespan in many regions around the globe;
5. modern dams have led to environmental destruction and silt buildups that require that they be torn down at great cost,[2] over-irrigation has

---

[1] www.vaxa.com/library/oldbugs.cfm.

[2] For a discussion of the problem of dam removal, see Mecoy (2002), while the problems of water scarcity are briefly examined by McCarthy (2003).

led to the abandonment of as much as 20 percent of previously arable land because of salinity buildup;[3] and overuse along with pollution has led to depletion of water resources that have made once-abundant fresh water the scarcest of all commodities on Earth;

6. the huge number of personal cars and trucks – now estimated at close to 800 million and predicted to reach one billion worldwide by 2020[4] – has led to congestion, sprawl, pollution, and social isolation; and

7. our great discoveries in physics and biology have led, among other things, to mass weaponry that has killed tens of millions of humans over the past century and continues to threaten to destroy much of the Earth's inhabitants.

Only literature, music, and art seem to be to be largely free of negative consequences, but they have contributed to an explosion of information that some believe is psychologically deleterious (Hall and Walton, 2004). Moreover, even the artistic and scientific endeavors may be reaching a state of diminishing returns, as reflected in the so-called "end of science"(Horgan, 1996), the absence of a Picasso or Beethoven or Proust in the past half-century or more, and more mundanely by the ever-growing number of film sequels and remakes. Indeed, the dopaminergic mind may even be turning against itself, as some of its products may render less valuable dopaminergically mediated abilities in humans such as abstract reasoning and dopaminergic goals such as space exploration. For example, no human can match the chess-playing ability or the computational skills of a computer, and virtual and other entertainment mediums have become so sophisticated and immersive that the fantasy of exploring other galaxies is much more enthralling (and certainly more cost-effective) to the average person than the reality of watching a robotic vehicle travel a few inches a day on Mars.

Until the latter half of the twentieth century, the greatest downside of our dopaminergic minds would unquestionably have been the seemingly endless progression of wars that were fought over abstract religious and political ideology, which claimed over 100 million lives in the twentieth century and regularly continue to claim over one million lives per year, along with many times that number of refugees. However, it can be argued that the greatest damage caused by the dopaminergic mind in the future will be to our environment (including other species) and to our own mental health. I believe that these twin perils – even more than the continual warfare that is seemingly tolerated by much, if not most, of the

---

[3] http://usinfo.state.gov/products/pubs/desertific/water.htm.

[4] http://online.wsj.com/public/article/SB114487401909024337-
ouwLdesvUMPaejrsk_WhxkaZzNU_20060516.html?mod=tff_main_tff_top.

world – will eventually force a reassessment of our traditional ideals and a correction of the current neurochemical imbalance favoring dopaminergic activity.

It has only been since the end of World War II that the tremendous negative impact of humans on the environment has been acknowledged and studied. While human-led extinctions of other species date back to prehistoric times, the rate of destruction of our fellow animal and plant species is occurring at an unprecedented rate because of deforestation, pollution, invasion of alien species, and climate change. It is estimated that extinctions are greater now than at any time following the emergence of complex organisms about 500 million years ago and that they dwarf the mass extinctions of dinosaurs and other creatures during the Cretaceous era. Indeed, the recent rate of extinction has been estimated at over 100,000 times what would have occurred naturally in the absence of humans (Leakey and Lewin, 1995). Until recently, the notion that humans could actually alter the Earth's climate would have been considered absurd, but that view is now widely accepted due to the global warming caused by a dramatic increase in fossil fuel consumption and the release of other gases such as methane that create a carbon-dioxide umbrella akin to a planetary greenhouse (Bowen, 2005). Per capita energy use in the United States has increased by almost 100 percent from the 1950s to the present (Zhou, 1998), and although the per capita rise has not been nearly as great in recent decades, total energy use continues to show major climbs due to a rapid increase in population – from 150 to 280 million in the United States and from two and a half to over six billion worldwide during the past half-century. Modest efficiencies in fuel use and household energy consumption have been offset by such factors as longer commuting and transportation distances, as huge amounts of energy are now required to move raw materials, consumer and other finished products, and even people to all parts of the globe. Global warming, which has already reached 4–5°C in the polar regions over the past one hundred years and will probably reach that same level over the entire planet by 2100, is obviously the single-greatest catastrophic threat to the current environment. In addition to the impending doom for thousands more animal and plant species, human-induced global warming and other environmental degradation will increasingly lead to large-scale human death and suffering, including the likelihood of hundreds of millions of environmental refugees fleeing expanding deserts and submerging coastlines.

The second important issue is whether the mental health of our society is threatened by an excessive amount of dopamine in our brains due to the constellation of societal pressures (e.g. constant change, reduced sleep amount and quality, extreme competitiveness, mental stress) created by modern industrialized societies. Just as it was once thought that human activity could not dramatically alter the world's climate, so would it have once seemed unlikely that modern societies and lifestyles could produce widespread and serious neurochemical imbalances and associated impairments in brain function. However, it is now increasingly recognized that hyperdopaminergic illnesses such as autism, bipolar disorder, and schizophrenia are either much more prevalent or more severe than in nonindustrialized societies and communities (Previc, 2007; Sanua, 1983; Sartorius et al., 1986). As noted in Chapter 4, the hyperdopaminergic disorders can be directly affected by the psychological stress of modern societies, which depletes the brain of serotonin and norepinephrine and reduces their inhibition of dopamine, leading to elevated levels of the latter. Depression and anxiety disorders have also skyrocketed, with a tripling of the number of adults who took antidepressants from 1994 to 2002 (Vedantum, 2004), despite the fact that even newer antidepressants are only effective in a minority of the population and can cause many negative side effects such as weight gain and sexual dysfunction (Moncrieff et al., 2004). Depression alone was estimated to cost the United States over $80 billion in 2000, and autism is predicted to cost the United States economy $400 billion by 2016.[5] But, these figures may represent only the tip of the iceberg in that once-rare autoimmune disorders and even the entire syndrome known as the "metabolic syndrome" (hypertension, diabetes, obesity) are all affected by serotonergic and noradrenergic depletion as levels of stress increase in our society (Muldoon et al., 2006). Excessive dopamine per se is not necessarily a direct cause of the serotonergic-deficiency syndromes, but because serotoninergic inputs normally inhibit dopamine in a variety of brain areas (see Chapters 2 and 4), the hyperdopaminergic and serotonin-deficiency syndromes may both be manifestations of the same psychological pressures in modern industrialized societies.

It would be convenient if we could deal with our looming environmental crises merely by inventing some new technologies and thereby avoiding a change to our core values and lifestyles. Indeed, there are environmental scientists who believe that future technologies can dramatically reduce global energy consumption by 90 percent without affecting the standard of living in industrialized countries, and even some environmental psychologists (Stern, 2000) believe that technological innovation may be superior to individual behavioral change in this regard. It is highly doubtful, however, that a technological change

[5] Dr. Cathy Pratt, Chairman of the Board of the Autism Society of America, *South Texas Autism Summit*, January 14, 2006.

alone could even begin to stave off climate change and other impending ecological disasters. For one, there are several factors that will offset any efficiencies gained from technological improvements (e.g. a gain in world population, increased natural solar absorption due to melting ice caps already underway, continuing release of heat from engines, machines, or devices using even renewable energy resources, greater cooling requirements due to higher global temperatures, increased development among less-industrialized nations, continued carbon emissions from agriculture etc.). Nor would any technological solution be without its own negative environmental impact; e.g. computers, cell phones, lasers, and other advanced electronic products require metals such as nickel and cadmium, the mining, refinement, and disposal of which may incur their own serious environmental damage. More importantly, there is little evidence that previous technological advances *in and of themselves* have done anything at all to protect us against the ravages of climate change, given that per capita energy consumption continues to rise in the United States, China, and most other nations. Despite more efficient cars, increases in the average number of cars per family, average number of miles traveled per car, and inefficient use (e.g. wasteful idling in increasingly prevalent traffic jams) have, along with the greater consumption of plastics and other synthetic products, actually increased the United States' per capita consumption of petroleum since 1970. And although our central air conditioning systems may be more efficient than in the past, increases in the average size of a new home offset many of these efficiencies. Moreover, even if we were to totally rely on renewable sources for energy production – an extremely unlikely prospect even in the next century – we would still face diminishing supplies of raw materials for everyday-use products of the industrial society.[6]

The inconceivability of a "technology-alone" solution does not necessarily require a return to a pre-industrial lifestyle, as some researchers believe is likely in what has been termed the "Olduvai Theory" (Duncan, 2005), so named after the gorge in which the remains of the first *Homo habilis* were found. But, as suggested by Capra (2003), Clark (1989), Howard (2000), Oskamp (2000) and others, any hope of avoiding environmental calamity in the future will require a *profound behavioral change* in which individuals turn away from conspicuous consumption, individual transport, and throw-away convenience to a lifestyle that reduces consumption, relies on energy-efficient mass

---

[6] For example, less than one-third of the United States' estimated copper reserves remain in the earth (Greenfieldboyce, 2006).

transportation, and emphasizes recycling and environmental sustainability (circular output) rather than the continuous production of new items (linear output). In addition, the human population growth that has almost continuously occurred for the past dozen or so millennia must be supplanted by decades if not centuries of population decline, with two billion humans considered by many experts as the maximum number possible to sustain a post-industrial society permanently depleted of cheap resources (Duncan, 2005). Just to return to the human environmental footprint of 1950 – or prior to the onset of runaway global warming, when the global population was 3.5 billion and per capita energy consumption in the United States was less than half that of today – would require a veritable revolution in the modern lifestyle. In the end, individual pro-environmental behavior may not merely be a "personal virtue," to quote former Vice-President Dick Cheney,[7] but rather a necessity to preserve the health of the planet.

The same anti-technological arguments hold true for the increase in hyperdopaminergic disorders such as autism. One might hope that some genetic breakthrough or new pharmacological treatment might stem the increase in autism and related disorders. However, gene therapy has proven risky and mostly unsuccessful even when a disorder is caused by a single gene with high penetrance, which is definitely not the case in autism and the five other major hyperdopaminergic disorders reviewed in Chapter 4. And, pharmacological treatments (mostly anti-dopaminergic ones) have heretofore proven of limited value in treating the social deficits found in the hyperdopaminergic disorders, particularly autism (Previc, 2007), just as anti-depressants are only marginally effective despite their widespread use (Moncrieff et al., 2004). It may, therefore, prove more advantageous (even necessary) to reduce the prevalence of hyperdopaminergic disorders by changing behavior (e.g. earlier child-rearing, reduced stress, increased leisure and social activity, more sleep etc.) that is known to elevate dopamine and increase the risk of autism in offspring rather than to bank on pie-in-the-sky technological remedies.

7.2   Tempering the dopaminergic mind

7.2.1   Altering dopamine with individual behavior

There is hardly a more ubiquitous expression than "You can't change human nature." It is true that destructive human-initiated behaviors such as wars, torture, greed and inequity, and environmental destruction

---

[7] www.commondreams.org/headlines01/0502–01.htm.

have been ongoing or even accelerating since the dawn of the ancient civilizations and perhaps even before. Yet, there are also pacifists to be found among warriors, selfless individuals among the greedy, and heroic efforts at environmental sustainability in the midst of massive environmental abuse, and there were once even entire civilizations that were relatively peaceful and environmentally benign (Taylor, 2005). Moreover, what is considered "human nature" depends to a great extent on gender, as certain negative dopaminergic traits such as violence are much less prevalent among women than men, whereas the reverse is the case for nondopaminergic traits such as social nurturance and cooperation. Finally, as described elsewhere in this book, the dopaminergic content of the human brain – the leading contributor to our abstract intelligence, exploratory drive, urge to control and conquer, religious impulses, obsession with achieving and creating things, acquisition of incentive rewards such as money, information and power, and choice of social domination over affiliation – is hardly a permanent or even static neurochemical phenomenon, as it was largely created by epigenetic forces. Hence, the notion that one "cannot change human nature" is less scientific fact than a polemical defense of the status quo in society (typically made by those who have the most financially to lose by change).

On the other hand, it will not be easy to reduce the dopaminergic content of the human brain, because the dopaminergic mind was forged by many environmental and societal adaptations and is the dominant side of our psyches. Four major obstacles to restoring balance to the dopaminergic mind through individual behavior are:

1. dopaminergic behavior is highly addictive psychologically;
2. dopamine promotes delusional behavior and, in a milder sense, rationalization and denial;
3. dopamine is fundamental to masculine behavior and is relatively higher in the very gender that dominates the political and economic life of most societies; and
4. dopamine is, to varying degrees, highly adaptive in dealing with the psychological stresses and uncertainties of modern life.

As reviewed in Chapter 4, dopamine is the most highly involved brain chemical involved in addictive behavior, including such diverse behaviors as substance abuse, sexual addictions, and gambling. Overactive dopamine systems may contribute to all forms of obsessive-compulsive activity and even obsessive-compulsive spectrum disorders such as video-game playing and workaholism that are now rampant in modern societies. Merely treating the overt addictions without reducing the underlying dopaminergic drive may not be all that productive; for example, a replacement of dopaminergic religious obsessions by sexual and other nonreligious obsessions has occurred in increasingly secular societies, without a fundamental reduction in the obsessive tendency itself (Previc, 2006). On the other hand, changing our fast-paced, addictive hyperdopaminergic lifestyle too quickly could lead to rampant boredom and even anxiety (Taylor, 2005). Moreover, even if an individual was inclined to accept a much lower-paying job (or any job at all) so as to increase badly needed leisure time and a slower lifestyle, it may be necessary to avoid such a transition for legitimate economic reasons (e.g. need to provide for ever-increasing college costs or to maintain adequate healthcare coverage).

The second serious impediment to changing dopamine levels is that dopamine is the neurochemical most likely to promote delusional behavior. For example, ingestion of dopaminergic agonists and stimulants such as amphetamine and cocaine can lead delusions of grandeur – the belief that one can inordinately understand and control external events – which is a residue of the role of dopamine in perceiving associations and predicting/controlling goal-relevant stimuli in extrapersonal space (see Chapter 3). Similarly, lack of insight and even denial is a common problem in treating substance-abuse, gambling, and other hyperdopaminergic obsessive-compulsive spectrum disorders and, as already reviewed, delusional traits are commonly found in societal leaders who frequently manifest strong dopaminergic personalities. Indeed, the hyperdopaminergic behaviors considered "normal" by a highly dopaminergic society may, like all other addictions, be most subject to denial. For example, a hunter-gatherer would probably consider a >60-hour workweek insane (or at least manic), yet it would be considered normal among executives in the United States and even worthy of social praise in some circles.[8] Even the efforts of many scientists to find seemingly impossible cures or uncover ultimate truths such as the origins of the universe may be considered mildly delusional and symptomatic of the failure of the dopamine-rich left hemisphere to acknowledge randomness and limits to our knowledge. The empirical data concerning these ultimate questions are currently very sparse and are likely to remain so, and scientific hypothesis-testing in a poor empirical environment essentially begins to resemble the quasi-superstitious (even delusional) behavior of the dopamine-rich "rational" left hemisphere in its fruitless quest to find predictability in a random

---

[8] Indeed, the average workweek in many lucrative professions in the United States is now greater than sixty hours: www.law.yale.edu/documents/pdf/CDO_Public/cdo-billable_hour.pdf.

environment (Wolford *et al.*, 2000). The delusional side of science is humorously captured in the famous quote about astrophysicists: "always in error, never in doubt" (Stenger, 2002).[9]

The third reason why it will be difficult to alter dopamine levels in at least half of the human race is that, aside from any increases throughout evolution and history, dopamine is naturally higher in males due to the testosterone–dopamine connection. As noted in Chapter 3, the drives to achieve, explore, analyze, and control are classic dopaminergic traits and, as discussed elsewhere, are more prominent in the dopamine-rich hemisphere and also in males because of the testosterone–dopamine link. Capra (2003), Primavesi (1991), Taylor (2005), and even Brown (2003, in the fictional *Da Vinci Code*) are but a few of the many recent authors who have argued that, for societal and environmental reasons, the dominant male behavioral style must be tempered by its fusion with the more nurturant female mode. However, this will not be easy given that men, or at least masculine behavioral traits (competitiveness and aggressiveness), dominate the political life of most nations on Earth, with even females who achieve leadership roles in society tending to acquire such traits. It is especially difficult for feminine values to triumph when women in industrialized countries who choose raising families (nurturing) over careers (competitiveness) for a large part of their working lives end up limiting their access to economic and political power. Not surprisingly, predominantly female occupations such as teaching, counseling, and nursing tend to be less lucrative and politically favored than corresponding male-dominated professions (Padavic and Reskin, 2003).

A final reason why reducing our dopaminergic content may prove extremely difficult is because high dopamine levels, while potentially very dangerous, are also highly adaptive in uncertain environments. Dopamine allows for active coping, in which we continue to engage in behavior (e.g. overachievement, extra work hours) merely to avoid negative consequences (e.g. job failure, family financial stress). Active coping styles in dopaminergic-rich individuals greatly contribute to what has been termed a "hardy" personality, which is especially invaluable in extreme psychological conditions such as wartime (Previc, 2004). Without such active coping, humans as well as other species tend to succumb to what is known

---

[9] As reviewed in Chapter 6, Newton with his biblical code-breaking and alchemy and Einstein with his search for the supposed "Holy Grail" of physics – the unified field theory – may have wasted much of their scientific lives on what were clearly or arguably unattainable discoveries. A more direct link between scientific and clinical delusion is the case of the Nobel Prize-winning mathematician John Nash, who is believed to have precipitated his longstanding psychosis (associated with elevated dopamine) by tackling the still-unsolved Riemann Hypothesis (Nasar, 1998).

as "learned helplessness" and psychological depression (Seligman, 1975). Uncertainty is arguably the single-greatest threat to the mental health of the average person, yet individuals in the industrialized societies in particular have become increasingly less stable due to global economic forces, reduced job security, heightened societal mobility, increasing marital failures, etc. To reduce the uncertainty found in modern industrial societies and thereby allow nondopaminergic lifestyles to flourish would mean, among other things, a roll-back of powerful economic forces such as globalization and rapid technological obsolescence.

Hence, changing individual behavior to reduce the negative consequences of the dopaminergic mind, however necessary, will be extremely difficult to achieve without concomitant changes in society.

7.2.2   Knocking down the pillars of the hyperdopaminergic society

The relinquishing of the dopaminergic imperative will require the replacement or at least tempering of several societal ideals and institutions that have dominated human history and have helped to build great civilizations and understand and control nature. The six major pillars of the hyperdopaminergic society to be reviewed here are masculine dominance, the drive to explore, the pursuit of wealth, the glorification of military conquest, a belief in the omnipotence of technology, and the predominance of anthropocentric and hierarchical religious institutions. In addition to requiring a major change in societal values, toppling these pillars will require overcoming powerful economic interests, including energy and automobile manufacturers, drug companies, and, of course, the well-known military-industrial complex.[10] Despite their strong underpinnings, however, these pillars are not invulnerable, as a turning away from these ideals is already beginning to occur in many industrial societies (see Taylor, 2005).

As just noted, one of the most common themes highlighted by those who have argued for fundamental societal change is the empowerment of women politically, socially, and economically. With the advent of technologically advanced and stratified societies, masculine domination (e.g. conquest, control) replaced the more egalitarian gender relations of hunter-gatherer societies, and this change was associated with the rise of patriarchal control and lineage, hierarchical male-dominated religions, and a more exploitive attitude toward nature (Primavesi, 1991; Taylor, 2005). Whereas technological prowess has had, especially in the past

---

[10] This term gained prominence in President Eisenhower's farewell address to the United States in 1961: http://coursesa.matrix msu.edu/~hst306/documents/indust.html.

century or so, masculine connotations (Oldenziel, 2004) – indeed, 90 percent of all engineers in the United States were males in a recent survey (National Science Foundation, 1999) – nature and environmental sustainability tend to connote femininity, as in the phrases "Mother Earth," "Mother Nature," and "Gaia" (the "Earth Goddess") (Primavesi, 1991). However, mere empowerment of women does not ensure the relinquishing of the dopaminergic imperative, since many famous and highly dopaminergic female leaders (e.g. Indira Gandhi, Golda Meir, Margaret Thatcher) have shown themselves more than willing to wage war and/or cut social services to the poor. Indeed, the recent epidemic of autism in the industrialized societies – which can be partly traced to high maternal dopamine levels, especially among the most educated classes – suggests that female professionals who possess high achievement-orientation and other traditionally male traits may have substantially elevated their dopamine levels (Previc, 2007). Hence, reducing dopamine levels in both men and women will require more than just the empowerment and masculinization of women; rather, it will require the restoration of more traditional female values of nurturance and affiliation across society as a whole.

As a testament to the strength of the second dopaminergic pillar, the human experience has been associated with great migrations, explorations, and/or expeditions since the dawn of *Homo habilis*. In the more recent historical era, most of these journeys and explorations have been associated with dopaminergic goals (e.g. militaristic, commercial, scientific, and religious). However, exploration may no longer constitute the idealized pursuit it was once was, for two major reasons. For one, all regions of the Earth with the exception of Antarctica have been populated and exploited. Second, the cost and risks of interplanetary space exploration are simply mind-boggling when weighed against its potential benefits. Although our dopaminergic minds may find it inspiring to look upward into the heavens, from a biological standpoint our solar system is a vast desert except for our lonely but beautiful and vibrant earthly oasis. One might find inhabitable planets in nearby solar systems, but the distances involved are staggering – traveling at the incredible speed of 1,000,000 kilometers per hour, it would take almost 5,000 years to reach Proxima Centauri, our nearest galactic neighbor. Even travel to a nearby planet like Mars poses enormous hazards from a biological standpoint – solar radiation, toxic atmosphere, a greater-than-nine-month confinement, potentially insufficient caloric intake, inability to overcome the Martian gravitational field on departure etc. – and is hence extremely unlikely to be conducted within our lifetimes (Easterbrook, 2004). Also, the tangible benefits to humankind from such travel are difficult to quantify, unlike the great explorations of the past. Even if Earth were hit by an asteroid or some other cosmic catastrophe, humans would still have an infinitely better chance of surviving for hundreds of years underground on Earth than on other planets and moons with poisonous atmospheres and no water, organic matter, or easily exploitable resources.

The sanctity of the third dopaminergic pillar – the pursuit of wealth – must also be challenged. The pursuit of wealth basically represents an addiction to secondary incentives such as money, power, and status, since basic needs can be provided with but a fraction of the per capita economic output of developed nations. As noted earlier in this chapter, wealthier nations create a highly disproportionate amount of environmental damage in the form of carbon emissions, air pollution, pesticide use, hazardous waste etc. Moreover, individuals in many of the wealthy nations also have little leisure time, greater psychosocial stress, and greater sleep deprivation. The pursuit of wealth and conspicuous consumption has reached almost bizarre extremes in present-day American society, as reflected in the infamous line by the corporate raider Gordon Gekko in the movie *Wall Street* that "greed is good"[11] and the belief of one national commentator that "global warming is great" because it creates "the investment opportunity of a lifetime."[12] The United Nations and other organizations have begun moving away from gross domestic product as a measure of quality of life and are now focusing on measures such as the Human Development Index, which focuses on such items as literacy and lifespan as well as economic output. Because gross domestic product is a component of the composite index as well as an independent influence on its non-economic measures such as literacy and lifespan, a fairly high correlation exists between the two measures (typically 0.8–0.9). However, the correlation is not nearly as good for the nonindustrialized countries, some of whom rank high on the Human Development Index but below average on per capita gross domestic product, thereby demonstrating that a high per capita economic output is by no means a requisite for a high quality of life. If the Human Development Index were to include such items as amount of leisure time, commuting time, and psychosocial indicators such as divorce, then it would unquestionably be even less well (if not negatively) correlated with per capita economic output. Furthermore, correlations between per

[11] The "greed is good" statement was based on a real speech by convicted financier Ivan Boesky at the University of California at Berkley: http://en.wikipedia.org/wiki/Gordon_Gekko.
[12] From a National Review online article by James Robbins: http://article.nationalreview.com/?q=ZTJmNWI4N2Y2NTBmY2E3Z7IzZjcxM2IzM2ZjNjRkYWI.

capita economic output and environmental measures are very negative, indicating that the richest nations of the world are producing the greatest worldwide environmental damage. For example, one measure known as the "ecological footprint" correlates -0.43 with per capita gross domestic product for the industrialized nations (Moldan *et al.*, 2004).

Glorification of military conquest as a "noble" enterprise is the fourth and perhaps most enduring component of the dopaminergic imperative, as such conquest combines the dopaminergic emphases on intense motivation, control, and distant goals with dopaminergically mediated male aggression. Intra-species violence is not limited to humans, as other primates and mammals may engage in violence and even killing to dominate their immediate social hierarchy or territory. Only humans, however, have traveled great distances to engage in wholesale slaughter for abstract ideals such as political ideology, religion, or even fame and glory. Military service, even in the industrialized nations, is arguably the most highly male-dominated of all professions, at least at the highest levels where combat experience is typically required for advancement. Also, male-dominated scientific fields such as physics and engineering are a highly valued component of military capabilities, as ~50 percent of all scientists and engineers in the United States government are employed by the Department of Defense (National Science Foundation, 1996). Conversely, women have often been the greatest victims of military conflict, in the form of rape and pillaging, and women have traditionally been the leaders of pacifist movements worldwide since the nineteenth century, including the Women's International League for Peace and Freedom, the Women's Peace League in Britain, The Federation for a New Fatherland in Weimar Germany, and the Northern Ireland Peace Movement. Deglorification and even elimination of large-scale military conflict cannot merely be considered an issue of individual behavior, but will in the end require the emergence of strong and truly representative international agencies and international forces capable of promoting and ensuring peaceful relations among nations and individuals.

The relentless pursuit of scientific discovery and the over-zealous faith in the power of technology comprise the fifth dopaminergic pillar that may have to be devalued in the future. As noted already, male-dominated or inspired scientific and technological advances have been intertwined with warfare throughout history, from the first bronze and iron weapons to the development of the atomic bomb and computers. Also, the scientific awakening in the Renaissance and Enlightenment went hand-in-hand with the belief by its progenitors (e.g. Descartes and Bacon) that science should help mankind to dominate and even enslave nature (Taylor, 2005). Of course, science is not inextricably tied to warfare and ecological profligacy, and it could in principle be redirected toward peaceful purposes and environmental sustainability. Scientific thinking indisputably offers a means for objectively viewing and defining a set of problems, and it thereby in principle promotes greater agreement and understanding among individuals and societies. And, we certainly require continued scientific and technological advances to help reverse the serious (even catastrophic) environmental damage of the past centuries. Yet, the belief in the omnipotence of science and technology has reached almost dogmatic status in the wealthier nations over the past century. Scientific and technological progress is usually viewed as an arrow or a train moving in one direction, but even technophiles can no longer avoid turning round and confronting the detritus left behind along the path of scientific progress. Have our scientific pursuits and technological achievements promoted peace among individuals and nations, have they been beneficial to other species and the planet's health in general, have they reduced worldwide energy consumption, have they eliminated disease and human suffering, have they improved our overall mental health and made us happier? Indeed, the bountiful technology that allows us to produce tens of thousands of consumer products every year has led to so many choices that people actually grow unhappy as a result (Schwartz, 2004), perhaps because of the constant dopaminergically mediated decision-making that such choices require. The overabundance of technical gadgetry in the industrialized nations has further led to a decrease in our level of social interaction, as is most dramatically illustrated by the rise of "hikikomori," a widespread disorder in Japan afflicting mainly young males who spend most of their early adulthood interacting with computers in isolation and unable to maintain social contact even with immediate family members (Watts, 2002). It should be noted that at least one major challenge to the dominance of scientific and technological pursuit has already occurred, in that there is a likelihood of diminishing future returns for most scientific endeavors (Horgan, 1996). Humans have gained a powerful knowledge of the cosmos, the atom, and the genome and, as this book has intimated, perhaps even the human brain itself may now largely be understood in terms of its evolution, lateralization, neurochemical organization, and principal functions. Moreover, what has heretofore remained unexplained – the origins of the cosmos, the existence of multiple, parallel universes, the existence of a unified force underlying all matter and energy – may well turn out to be beyond the ability of the human mind to comprehend and/or its technological creations to measure or recreate.

The final pillar of the hyperdopaminergic society – the domination by hierarchical religious institutions – has provided a major role in driving mainly men to shape the course of history, as during the colonial era when servants of the colonial powers such as the conquistadors fought as much for the glory of God as for their own glory and wealth. Primavesi (1991), Taylor (2005), and others have argued that the insertion of masculine dominance into hierarchical, anthropocentric religions has also shaped our exploitive attitudes toward the environment, especially in the Judeo-Christian tradition.[13] The many positive effects of the major religions throughout history have been tainted by their complicity in innumerable wars and conflicts and in some of the worst crimes of humanity, including the slaughter of thousands of persons in Spain alone during the Inquisition in the late Middle Ages, the holding of over 100,000 "witch" trials during that same period over other parts of Europe, and the reduction of the population of the Americas from over fifty million to but a few million due to disease and slaughter following the colonial/Christian occupation. Yet, hierarchical religious institutions represent the dopaminergic societal pillar beginning to show the most cracks in modern societies, for as science moves back the frontiers of the unknown the concept of a personal God is becoming steadily more alien across the industrialized world. With the exception of the United States, religion is currently considered as important by approximately one-third or less of every major industrialized nation in the world.[14] Science has even encroached onto the source of religious behavior itself, as we now know that religious behavior is, like many other uniquely human behaviors, a product of the dopaminergic pathways coursing through the ventromedial brain (especially in the left hemisphere) that are oriented toward distant space and time and divorced from peripersonal/self systems (Previc, 2006). Of course, science cannot replace the mystery of life entirely, but it is doubtful whether rigid, hierarchical religious institutions can thrive only on mysteries remaining at the cosmic and atomic levels. Relinquishing the dopaminergic imperative need not result in a complete abandonment of religious and spiritual belief but only of particular religious systems linked to masculinity, hierarchy, domination, ecological destruction, and conflict.

7.3   Toward a new consciousness

The devaluation of dopaminergic ideals, while necessary, will not in and of itself reduce the dopaminergic content of our brain to more benign

---

[13] One oft-cited biblical passage to illustrate the Judeo-Christian exploitive attitude is "Rule over the fish of the seas and the birds of the air and over the livestock, over all the earth and over all the creatures that move along the ground" (*Genesis 1:28*).

[14] www.religioustolerance.org/rel_impo.htm.

levels. Powerful steps must also be taken on a worldwide basis to reduce one of the prime reasons that high dopamine levels are so adaptive – to control aversive and uncertain events. A reduction in psychological stress and uncertainty is needed to move the neurochemical balance away from dopamine and toward norepinephrine and serotonin, which are ordinarily depleted by stress. In a world where basic needs are met and one's life is not torn asunder by job losses, loss of healthcare coverage, intense competition, divorce, or uprooting to new locales, the need for and obsession with constant striving would be mitigated. To achieve such a world, we must make our economies and societies more egalitarian, less work-obsessed, less achievement-driven, and more stable. This societal shift would mean increasing the scope and strength of the social safety net, reversing the trend toward globalization (in which jobs jump from one nation to another in search of lower wages and reduced environmental protection, even as the ecological costs of transporting goods increases), and more carefully limiting technological advances until their consequences have been more fully examined. It would also mean reversing the global warming and other environmental damage that has forced millions of humans into economic desperation, migration, psychological despair and that, without prompt action, could lead to the deaths of hundreds of millions, if not billions, due to starvation.

To our revved-up (hypomanic) dopaminergic brains, a return to a more balanced neurochemical existence might initially produce negative consequences such as boredom, but in time a quieting of the dopaminergic output could be passed on to future generations in the same way that elevated dopaminergic activity is currently being passed on. And, while exploration, scientific discovery, and technological invention might consequently slow, the major dramas of human existence – birth, love, marriage, children, aging, suffering, death – will continue to be experienced and painted by writers and artists using the words and canvases of future generations.

Relinquishing the dopaminergic imperative will lead to a post-dopaminergic consciousness in which the key traits of the dopaminergic mind (e.g. far-sightedness, restlessness, detachment, exploitiveness, intense motivation for secondary goals such as wealth or abstract ideas) will be balanced or even replaced by their opposites (see Figure 7.1). It will involve creating a world in which the restlessness created by a future orientation is replaced by the contentedness of a present orientation, in which masculine values no longer dominate over feminine ones, where enhancing mental health becomes more important than the pursuit of wealth, where sustainability is more highly valued than exploitability, where preserving the lushness of Earth is more important than exploring

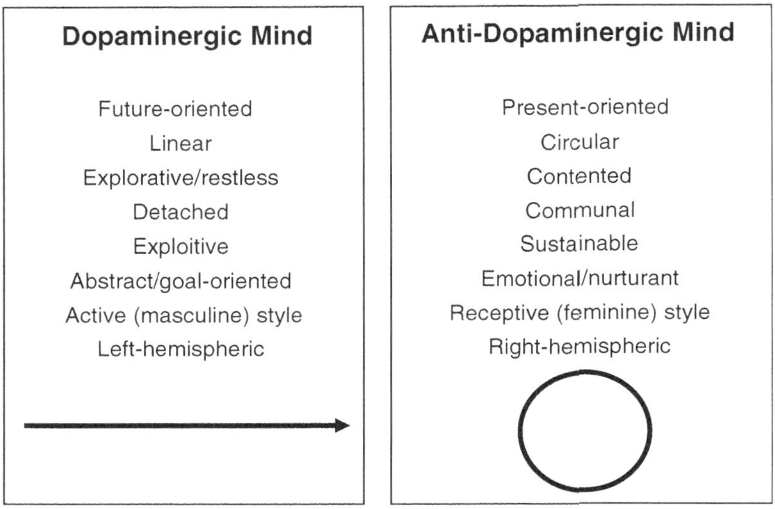

Dopaminergic Mind	Anti-Dopaminergic Mind
Future-oriented	Present-oriented
Linear	Circular
Explorative/restless	Contented
Detached	Communal
Exploitive	Sustainable
Abstract/goal-oriented	Emotional/nurturant
Active (masculine) style	Receptive (feminine) style
Left-hemispheric	Right-hemispheric

Figure 7.1 Restoring balance to the dopaminergic mind.

the desert of space, where a greater understanding and respect for our natural world and cosmos and their inherent mysteries replace both the ignorance of dogmatic religion and the arrogance of unrestrained science, where peacemakers are more valued than warriors, where nurturant emotional relationships are as important as abstract ideas and pursuits, where left-hemispheric activity no longer prevails over right-hemispheric activity, where embracing "the circle of life" is more important than following a linear "path to progress." It will not be easy for us to transition to a value system that predominated before the dopaminergic imperative reigned supreme, even as we attempt to retain much of our modern economic and technological infrastructure. However, the key message of this book is that *the dopaminergic imperative can indeed be relinquished because it is not primarily part of our genetic inheritance*. Just as our dopaminergic minds were forged by ecological forces and pressures such as thermal adaptation, diet, and exercise and later by the societal adoption of the dopaminergic imperative, so, too, can our minds be returned to a healthier balance over successive generations by altering our ideals and societies.

References

Abbruzzese, M., Ferri, S., and Scarone, S. (1995). Wisconsin Card Sorting Test performance in obsessive-compulsive disorder: no evidence for involvement of dorsolateral prefrontal cortex. *Psychiatry Research*, 58, 37–43.

Abi-Dargham A., and Moore, H. (2003). Prefrontal DA transmission at D1 receptors and the pathology of schizophrenia. *Neuroscientist*, 9, 404–416.

Abraham, A., Windmann, S., Siefen, R., Daum, I., and Gunturkun, O. (2006). Creative thinking in adolescents with attention deficit hyperactivity disorder (ADHD). *Child Neuropyschology*, 12, 111–123.

Abwender, D. A., Trinidad, K. S., Jones, K. R., Como, P. G., Hymes, E., and Kurlan, R. (1998). Features resembling Tourette's syndrome in developmental stutterers. *Brain and Language*, 62, 455–464.

Adler, C. M., McDonough-Ryan, P., Sax, K. W., Holland, S. K., Arndt, S., and Strakowski, S. M. (2000). FMRI of neuronal activation with symptom provocation in unmedicated patients with obsessive compulsive disorder. *Journal of Psychiatric Research*, 34, 317–324.

Al-Absi, M., Bongard, S., Buchanan, T., Pincomb, G. A., Licinio, J., and Lovallo, W. R. (1997). Cardiovascular and neuroendocrine adjustment to public speaking and mental arithmetic stressors. *Psychophysiology*, 34, 266–275.

Al-Adawi, S., Dawe, G. S., and Al-Hussaini, A. A. (2000). Aboulia: neurobehavioural dysfunction of dopaminergic system? *Medical Hypotheses*, 54, 523–530.

Alcaro, A., Huber, R., and Panksepp, J. (2005). Behavioral functions of the mesolimbic dopaminergic system: an affective neuroethological perspective. *Brain Research Reviews*, 56, 233–321.

Aldridge, J., Berridge, K., Herman, M., and Zimmer, L. (1993). Neuronal coding of serial order, syntax of grooming in the neostriatum. *Psychological Science*, 4, 391–395.

Aleman, A., Kahn, R. S., and Selten, J. P. (2003). Sex differences in the risk of schizophrenia: evidence from meta-analysis. *Archives of General Psychiatry*, 60, 565–571.

Alias, A. G. (2000). Schizotypy and leadership: a contrasting model for deficit symptoms, and a possible therapeutic role of sex hormones. *Medical Hypotheses*, 54, 537–552.

Alonso, L. C., and Rosenfield, R. L. (2003). Molecular genetic and endocrine mechanisms of hair growth. *Hormone Research*, 60, 1–13.

Ambrose, S. H. (1998). Late Pleistocene human population bottlenecks, volcanic winter, and differentiation of modern humans. *Journal of Human Evolution*, 34, 623–651.

Amos, J. (2 Dec 2003). Neanderthal "face" found in Loire. *BBC News World Edition Online* (http://news.bbc.co.uk/2/hi/science/nature/3256228.stm).

Andrew, J. M. (1981). Parietal laterality and violence. *International Journal of Neuroscience*, 12, 7–14.

Angst, J., Gamma, A., Endrass, J., Hantouche, E., Goodwin, R., Ajdacic, V., Eich, D., and Rossler, W. (2005). Obsessive-compulsive syndromes and disorders: significance of comorbidity with bipolar and anxiety syndromes. *European Archives of Psychiatry and Clinical Neuroscience*, 255, 65–71.

Anisman, H., and Zacharko, R. M. (1986). Behavioral and neurochemical consequences associated with stressors. *Annals of the New York Academy of Sciences*, 467, 205–225.

Annett, M. (1985). *Left, Right, Hand and Brain: the Right Shift Theory*. London: Erlbaum.

Arbisi, P. A., Depue, R. A., Krauss, S., Spoont, M. R., Leon, A., Ainsworth, B., and Muir, R. (1994). Heat-loss response to a thermal challenge in seasonal affective disorder. *Psychiatry Research*, 52, 199–214.

Archer, T., and Fredriksson, A. (1992). Functional changes implicating dopaminergic systems following perinatal treatments. *Developmental Pharmacological Therapy*, 18, 201–222.

Arensburg, B., Schepartz, L. A., Tillier, A. M., Vandermeersch, B., and Rak, Y. (1990). A reappraisal of the anatomical basis for speech in Middle Palaeolithic hominids. *American Journal of Physical Anthropology*, 83, 137–146.

Arnold, L. M. (2003). Gender differences in bipolar disorder. *Psychiatric Clinics of North America*, 26, 595–620.

Arnulf, I., Bonnet, A.-M., Damier, P., Bejjani, B.-P., Seilhean, D., Derenne, J.-P., and Agid, Y. (2000). Hallucinations, REM sleep, and Parkinson's disease. *Neurology*, 55, 281–288.

Arsuaga, J. L. (2003). *The Neanderthal's Necklace: in Search of the First Thinkers*. New York: Wiley & Sons.

Ashford, N. A., and Miller, C. S. (1991). *Chemical Exposures: Low Levels and High Stakes*. New York: Van Nostrand Reinhold.

Bäckman, L., and Farde, L. (2001). Dopamine and cognitive functioning: brain imaging findings in Huntington's disease and normal aging. *Scandinavian Journal of Psychology*, 42, 287–296.

Bäckman, L., Nyberg, L., Lindenberger, U., Li, S. C., and Farde, L. (2006). The correlative triad among aging, dopamine, and cognition: current status and future prospects. *Neuroscience and Biobehavioral Reviews*, 30, 791–807.

Baldo, B. A., and Kelley, A. E. (2007). Discrete neurochemical coding of distinguishable motivational processes: insights from nucleus accumbens control of feeding. *Psychopharmacology*, 191, 439–459.

Ball, D., Hill, L., Eley, T. C., Chorney, M. J., Chorney, K., Thompson, L. A., Detterman, D. K., Benbow, C., Lubinski, D., Owen, M., McGuffin, P., and Plomin, R. (1998). Dopamine markers and cognitive ability. *Neuroreport*, 26, 347–349.

Ball, W. W. Rouse (1908). *A Short Account of the History of Mathematics* (4th edn). Mineola: Dover Publications.

Bansal, S. A., Lee, L. A., and Woclf, P. D. (1981). Dopaminergic stimulation and inhibition of growth hormone secretion in normal man: studies of the pharmacological specificity. *Journal of Clinical Endocrinology and Metabolism*, 53, 1273–1277.

Barcelo, F., Sanz, M., Molina, V., and Rubia, F. J. (1997). The Wisconsin Card Sorting Test and the assessment of frontal function: a validation study with event-related potentials. *Neuropsychologia*, 35, 399–408.

Bardo, M. T., Donohew, R. L., and Harrington, N. G. (1996). Psychobiology of novelty seeking and drug seeking behavior. *Behavioral and Brain Research*, 77, 23–43.

Barrett, A. M., and Eslinger, P. J. (2007). Amantadine for adynamic speech: possible benefit for aphasia? *American Journal of Physical Medicine and Rehabilitation*, 86, 605–612.

Barros, R. C., Branco, L. G., and Carnio, E. C. (2004). Evidence for thermoregulation by dopamine D1 and D2 receptors in the anteroventral preoptic region during normoxia and hypoxia. *Brain Research*, 1030, 165–171.

Bates, T., and Stough, C. (1998). Improved reaction time method, information processing speed, and intelligence. *Intelligence*, 26, 53–62.

Batson, C. D., and Ventis, W. L. (1982). *The Religious Experience: a Social-Psychological Perspective*. New York: Oxford University Press.

Bauer, M., and Pfennig, A. (2005). Epidemiology of bipolar disorders. *Epilepsia*, 46 (Suppl. 4), 8–13.

Bechara, A., Tranel, D., and Damasio, H. (2000). Characterization of the decision-making deficit of patients with ventromedial prefrontal cortex lesions. *Brain*, 123, 2189–2202.

Behar, D. M., Villems, R., Soodyall, H., Blue-Smith, J., Pereira, L., Metspalu, E., Scozzari, R., Makkan, H., Tzur, S., et al. (2008). The dawn of human matrilineal diversity. *American Journal of Human Genetics*, 82, 1130–1140.

Bellini, S., and Pratt, C. (2003). Indiana parent/family 2003 needs assessment survey summary (www.iidc.indiana.edu/irca/2003NeedsAssess.html).

Beninger, R. J. (1983). The role of dopamine in locomotor activity and learning. *Brain Research Reviews*, 6, 173–196.

Benkelfat, C., Nordahl, T. E., Semple, W. E., King, A. C., Murphy, D. L., and Cohen, R. M. (1990). Local cerebral glucose metabolic rates in obsessive-compulsive disorder. Patients treated with clomipramine. *Archives of General Psychiatry*, 47, 840–848.

Bentley, A. (1983). Personal and global futurity in Scottish and Swazi students. *Journal of Social Psychology*, 121, 223–229.

Benvenga, S., Lapa, D., and Trimarchi, F. (2003). Don't forget the thyroid in the etiology of psychoses. *American Journal of Medicine*, 115, 159–160.

Berger, B., Gaspar, P., and Verney, C. (1991). Dopaminergic innervation of the cerebral cortex: unexpected differences between rodents and primates. *Trends in Neurosciences*, 14, 21–27.

Berlinski, D. (2000). *Newton's Gift*. New York: Free Press.

Berman, K. F., and Weinberger, D. R. (1990). Lateralisation of cortical function during cognitive tasks: regional cerebral blood flow studies of normal individuals and patients with schizophrenia. *Journal of Neurology, Neurosurgery, and Psychiatry*, 53, 150–160.

Berrettini, W. H. (2000). Are schizophrenic and bipolar disorders related? A review of family and molecular studies. *Biological Psychiatry*, 48, 531–538.

Berridge, K. C., and Robinson, T. E. (1998). What is the role of dopamine in reward: hedonic impact, reward learning, or incentive salience? *Brain Research Reviews*, 28, 309–369.

Bick, P. A., and Kinsbourne, M. (1987). Auditory hallucinations and subvocal speech in schizophrenic patients. *American Journal of Psychiatry*, 144, 222–225.

Bickerton, D. (1995). *Language and Human Behavior*. Seattle: University of Washington Press.

Biederman, J., and Faraone, S. V. (2005). Attention-deficit hyperactivity disorder. *Lancet*, 366, 237–248.

Bjerknes, C. J. (2002). *Albert Einstein: the Incorrigible Plagiarist*. Downers Grove: XTX, Inc.

Blackburn, J. R., Pfaus, J. G., and Phillips, A. G. (1992). Dopamine functions in appetitive and defensive behaviours. *Progress in Neurobiology*, 39, 247–279.

Blanchard, R. J., Griebel, G., Guardiola-Lemaitre, B., Brush, M. M., Lee, J., and Blanchard, D. C. (1997). An ethopharmacological analysis of selective activation of 5-HT1A receptors: the mouse 5-HT1A syndrome. *Pharmacology, Biochemistry and Behavior*, 57, 897–908.

Blatteis, C. M., Billmeier, G. J., Jr., and Gilbert, T. M. (1974). Thermoregulation of phenylketonuric children. *Pediatric Research*, 8, 809–814.

Blum, D. (1997). The plunge of pleasure. *Psychology Today*, Sep/Oct 97 (www.psychologytoday.com/articles/pto-19970901–000033.html).

Blumberg, H. P., Stern, E., Martinez, D., Ricketts, S., de Asis, J., White, T., Epstein, J., Mcbride, P. A., Eidelberg, D., Kocsis, J. H., and Silbersweig, D. A. (2000). Increased anterior cingulate and caudate activity in bipolar mania. *Biological Psychiatry*, 48, 1045–1052.

Bogen, J. E., and Bogen, J. M. (1988). Creativity and the corpus callosum. *Psychiatric Clinics of North America*, 11, 293–301.

Borod, J, C., Bloom, R. L., Brickman, A. M., Nakhutina, L., and Curko, E. A. (2002). Emotional processing deficits in individuals with unilateral brain damage. *Applied Neuropsychology*, 9, 23–36.

Bortz, W. M. H. (1985). Physical exercise as an evolutionary force. *Journal of Human Evolution*, 14, 145–155.

Botzung, A., Denkova, E., and Manning, L. (2008). Experiencing past and future personal events: functional neuroimaging evidence on the neural bases of mental time travel. *Brain and Cognition*, 66, 202–212.

Bowen, M. (2005). *Thin Ice: Unlocking the Secrets of Climate in the World's Highest Mountains*. New York: Holt and Company.

Boyages, S. C., and Halpern, J. P. (1993). Endemic cretinism: toward a unifying hypothesis. *Thyroid*, 3, 59–69.

Boyd, A. E., Lebovitz H. E., and Pfeiffer, J. B. (1970). Stimulation of human growth-hormone secretion by L-DOPA. *New England Journal of Medicine*, 283, 1425–1429.

Bracha, H. S., Cabrera, F. J. Jr., Karson, C. N., and Bigelow, L. B. (1985). Lateralization of visual hallucinations in chronic schizophrenia. *Biological Psychiatry*, 20, 1132–1136.

Bradshaw, J. L., and Nettleton, N. C. (1981). The nature of hemispheric specialization in man. *Behavioral and Brain Sciences*, 4, 51–63.

Brady, J. P. (1991). The pharmacology of stuttering: a critical review. *American Journal of Psychiatry*, 148, 1309–1316.

Bramble, D. M., and Lieberman, D. E. (2004). Endurance running and the evolution of Homo. *Nature*, 432, 345–352.

Braun, C. M., Dumont, M., Duval, J., Hamel, I., and Godbout, L. (2003). Opposed left and right brain hemisphere contributions to sexual drive: a multiple lesion case analysis. *Behavioral Neurology*, 14, 55–61.

Braver, T. S., and Barch, D. M. (2002). A theory of cognitive control, aging cognition, and neuromodulation. *Neuroscience and Biobehavioral Reviews*, 26, 809–817.

Breier, A, Kestler, L., Adler, C., Elman, I., Wiesenfeld, N., Malhotra, A., and Pickar, D. (1998). Dopamine D2 receptor density and personal detachment in healthy subjects. *American Journal of Psychiatry*, 155, 1440–1442.

Brian, D. (1996). *Einstein: a Life*. New York: Wiley & Sons.

Britton, W. B., and Bootzin, R. R. (2004). Near-death experiences and the temporal lobe. *Psychological Science*, 15, 254–258.

Broadhurst, C. L., Cunnane, S. C., and Crawford, M. A. (1998). Rift Valley lake fish and shellfish provided brain-specific nutrition for early Homo. *British Journal of Nutrition*, 79, 3–21.

Brock, J. W., and Ashby, C. R., Jr. (1996). Evidence for genetically mediated dysfunction of the central dopaminergic system in the stargazer rat. *Psychopharmacology*, 123, 199–205.

Brown, A. S. (2006). Prenatal infection as a risk factor for schizophrenia. *Schizophrenia*, 32, 200–202.

Brown, D. (2003). *The Da Vinci Code*. New York: Doubleday.

Brown, R. M., Crane, A. M., and Goldman, P. S. (1979). Regional distribution of monoamines in the cerebral cortex and subcortical structures of the rhesus monkey: concentrations and in vivo synthesis rates. *Brain Research*, 168, 133–150.

Brown, S. (2000). Evolutionary models of music: from sexual selection to group selection. In N. S. Thompson and F. Tonneau (eds), *Perspectives in Ethology 13: Evolution, Culture, and Behavior* (pp. 221–281). New York: Plenum.

Brozoski, T. J., Brown, R. M., Rosvold, H. E., and Goldman, P. S. (1979). Cognitive deficit caused by regional depletion of dopamine in prefrontal cortex of rhesus monkey. *Science*, 205, 929–932.

Brugger, P., Regard, M., and Landis, T. (1991). Belief in extrasensory perception and illusory control: a replication. *Journal of Psychology*, 125, 501–502.

Brumback, R. A. (1988). Childhood depression and medically treatable learning disability. In D. L. Molfese and S. J. Segalowitz (eds), *Brain Lateralization in Children: Developmental Implications* (pp. 463–505). New York: Guilford.

Brunet, M., Beauvilain, A., Coppens, Y., Heintz, E., Moutaye, A. H. E., and Pilbeam, D. (1995). The first australopithecine 2,500 kilometers west of the Rift Valley (Chad). *Nature*, 378, 273–278.

Bryan, K. L. (1988). Assessment of language disorders after right hemisphere damage. *British Journal of Disorders of Communication*, 23, 111–125.

Buchanan, R. W., Kirkpatrick, B., Heinrichs, D. W., and Carpenter, W. T., Jr. (1990). Clinical correlates of the deficit syndrome of schizophrenia. *American Journal of Psychiatry*, 147, 290–294.

Buka, S. L., and Fan, A. P. (1999). Association of prenatal and perinatal complications with subsequent bipolar disorder and schizophrenia. *Schizophrenia Research*, 29, 113–119.

Bunney, W. E., and Bunney, B. G. (2000). Evidence for a compromised dorsolateral prefrontal cortical parallel circuit in schizophrenia. *Brain Research Brain Research Reviews*, 31, 138–146.

Burgard, P. (2000). Development of intelligence in early treated phenylketonuria. *European Journal of Pediatrics*, 159, S74–S79.

Burgdorf, J., and Panksepp, J. (2006). The neurobiology of positive emotions. *Neuroscience and Biobehavioral Reviews*, 30, 172–187.

Caldwell, J. C., and Caldwell, B. K. (2003). Was there a Neolithic mortality crisis? *Journal of Population Research*, 20, 153–168.

Cann, R. L., Stoneking, M., and Wilson, A. C. (1987). Mitochondrial DNA and human evolution. *Nature*, 325, 31–36.

Capra, F. (2003). *The Hidden Connections*. New York: Doubleday.

Carpenter, L. L., Mcdougle, C. J., Epperson, C. N., and Price, L. H. (1996). A risk-benefit assessment of drugs used in the management of obsessive-compulsive disorder. *Drug Safety*, 15, 116–134.

Carpenter, P. A., Just, M. A., and Shell, P. (1990). What one intelligence test measures: a theoretical account of the processing in the Raven Progressive Matrices Test. *Psychological Review*, 97, 404–431.

Carrier, D. R. (1984). The energetic paradox of human running and hominid evolution. *Current Anthropology*, 2, 483–495.

Carroll, S. B. (2003). Genetics and the making of *Homo sapiens. Nature*, 422, 849–857.

Carson, S. H., Peterson, J. B., and Higgins, D. M. (2003). Decreased latent inhibition is associated with increased creative achievement in high-functioning individuals. *Journal of Personality and Social Psychology*, 85, 499–506.

Castle, D. J. (2000). Women and schizophrenia: an epidemiological perspective. In D. J. Castle, J. Mcgrath and J. Kulkarni (eds), *Women and Schizophrenia* (pp. 19–31). Cambridge, UK: Cambridge University Press.

Castle, D. J., Deale, A., and Marks, I. M. (1995). Gender differences in obsessive compulsive disorder. *Australian and New Zealand Journal of Psychiatry*, 29, 114–117.

Centers For Disease Control, and Prevention (CDC). (2005). Mental health in the United States. Prevalence of diagnosis and medication treatment for attention-deficit/hyperactivity disorder–United States, 2003. *Morbidity and Mortality Weekly Report*, 54(34), 842–847.

Chaouloff, F. (1989). Physical exercise and brain monoamines: a review. *Acta Physiologica Scandanavica*, 137, 1–13.

Charette, F. (2006). High tech from Ancient Greece. *Nature*, 444, 551–552.

Cheyne, J. A., and Girard, T. A. (2004). Spatial characteristics of hallucinations associated with sleep paralysis. *Cognitive Neuropsychiatry*, 9, 281–300.

Chomsky, N. (1988). *Language and Problems of Knowledge: the Managua Lectures.* Cambridge, MA: MIT Press.

Chua S. E., and McKenna, P. J. (1995). Schizophrenia – a brain disease? A critical review of structural and functional cerebral abnormality in the disorder. *British Journal of Psychiatry*, 166, 563–582.

Chudasama, T., Nathwani, F., and Robbins, T. W. (2005). D-amphetamine remediates attentional performance in rats with dorsal prefrontal lesions. *Brain Research*, 158, 97–107.

Civelli, O. (2000). Molecular biology of the dopamine receptor subtypes. In *Neuropsychopharmacology: the Fourth Generation of Progress.* American College of Neuropsychopharmacology (www.acnp.org/content-32.html).

Clark, D., and Clark, S. P. H. (2001). *Newton's Tyranny: the Suppressed Scientific Discoveries of Stephen Gray and John Flamsteed.* New York: W. H. Freeman and Company.

Clark, M. E. (1989). *Ariadne's Thread: the Search for New Modes of Thinking.* New York: St. Martin's Press.

Cloninger, C. R., Svrakic, D. M., and Przybeck, T. R. (1993). A psychobiological model of temperament and character. *Archives of General Psychiatry*, 50, 975–990.

Coco, M. L., and Weiss, J. M. (2005). Neural substrates of coping behavior in the rat: possible importance of mesocorticolimbic dopamine system. *Behavioral Neuroscience*, 119, 429–445.

Cohen, M. S., Kosslyn, S. M., Breiter, H. C., Digirolamo, G. J., Thompson, W. L, Anderson, A. K., Brookheimer, S. Y., Rosen, B. R., and Belliveau, J. W. (1996). Changes in cortical activity during mental rotation. A mapping study using functional MRI. *Brain*, 119, 89–100.

Comings, D. E., and Blum, K. (2000). Reward deficiency syndrome: genetic aspects of behavioral disorders. *Progress in Brain Research*, 126, 325–341.

Comings, D. E., Wu, S., Chiu, C., Ring, R. H., Gade, R., Ahn, C., MacMurray, J. P., Dietz, G., and Muhleman, D. (1996). Polygenic inheritance of Tourette syndrome, stuttering, attention deficit hyperactivity, conduct, and oppositional defiant disorder: the additive and subtractive effect of the three dopaminergic genes–DRD2, D beta H, and DAT1. *American Journal of Medical Genetics*, 67, 264–288.

Como, P. G., Lamarsh, J., and O'Brien, K. A. (2005). Obsessive-compulsive disorder in Tourette's syndrome. *Advances in Neurology*, 96, 249–261.

Cooper, J. R., Bloom, F. E., and Roth, R. H. (2002). *The Biochemical Basis of Neuropharmacology* (8th edn). New York: Oxford University Press.

Coppens, Y. (1996). Brain, locomotion, diet, and culture: how a primate, by chance, became a man. In J.-P. Changeux and J. Chavaillon (eds), *Origins of the Human Brain* (pp. 104–112). Oxford: Clarendon.

Corballis, M. C. (1989). Laterality and human evolution. *Psychological Review*, 96, 492–505.

Corballis, M. C. (1992). On the evolution of language and generativity. *Cognition*, 44, 197–226.

Coren, S. (1992). *The Left-Hander Syndrome: the Causes and Consequences of Left-Handedness*. New York: Free Press.

Corin, M. S., Elizan, T. S., and Bender, M. B. (1972). Oculomotor function in patients with Parkinson's disease. *Journal of the Neurological Sciences*, 15, 251–265.

Corti, O., Hampe, C., Darios, F., Ibanez, P., Ruberg, M., and Brice, A. (2005). Parkinson's disease: from causes to mechanisms. *Comptes Rendus Biologies*, 328, 131–142.

Cox, B. (1979). Dopamine. In P. Lomax and E. Schonbaum (eds), *Body Temperature: Regulation, Drug Effects, and Therapeutic Implications* (pp. 231–255). New York: Dekker.

Crile, G. (1934). *Diseases Peculiar to Civilized Man*. New York: MacMillan.

Crockford, S. J. (2002). Commentary: thyroid hormone in Neandertal evolution: a natural or pathological role? *Geographical Review*, 92, 73–88.

Croen, L. A., Grether, J. K., and Selvin, S. (2002). Descriptive epidemiology of autism in a California population: who is at risk? *Journal of Autism and Developmental Disorders*, 32, 217–224.

Cropley, V. L., Fujita, M., Innis, R. B., and Nathan, P. J. (2006). Molecular imaging of the dopaminergic system and its association with human cognitive function. *Biological Psychiatry*, 59, 898–907.

Crow, T. J. (1973). Catecholamine-containing neurones and electrical self-stimulation: 2. A theoretical interpretation and some psychiatric implications. *Psychological Medicine*, 3, 66–73.

Crow, T. J. (2000). Schizophrenia as the price that homo sapiens pays for language: a resolution of the central paradox in the origin of the species. *Brain Research Brain Research Reviews*, 31, 118–129.

Cummings, J. L. (1995). Anatomic and behavioral aspects of frontal-subcortical circuits. *Annals of the New York Academy of Sciences*, 769, 1–13.

Cummings, J. L. (1997). Neuropsychiatric manifestations of right hemisphere lesions. *Brain and Language*, 57, 22–37.

Cutting, J. (1990). *The Right Hemisphere and Psychiatric Disorders*. Oxford: Oxford University Press.

Damsa, C., Bumb, A., Bianchi-Demicheli, F., Vidailhet, P., Sterck, R., Andreoli, A., and Beyenburg, S. (2004). Dopamine-dependent side effects of selective serotonin reuptake inhibitors: a clinical review. *Journal of Clinical Psychiatry*, 65, 1064–1068.

Daneman, M., and Merikle, P. M. (1996). Working memory and language comprehension: a meta-analysis. *Psychonomic Bulletin and Review*, 3, 422–433.

Davidson, L. L., and Heinrichs, R. W. (2003). Quantification of frontal and temporal lobe brain-imaging findings in schizophrenia: a meta-analysis. *Psychiatry Research*, 122, 69–87.

Davis, J. O., Phelps, J. A., and Bracha, H. S. (1995). Prenatal development of monozygotic twins and concordance for schizophrenia. *Schizophrenia Bulletin*, 21, 357–366.

Dawson, M., Soulières, I., Gernsbacher, M. A., and Mottron, L. (2007). The level and nature of autistic intelligence. *Psychological Science*, 18, 657–662.

de Almeida, R. M., Ferrari, P. F., Parmigiani, S., and Miczek, K. A. (2005). Escalated aggressive behavior: dopamine, serotonin and GABA. *European Journal of Pharmacology*, 526, 51–64.

de Bode, S., and Curtiss, S. (2000). Language after hemispherectomy. *Brain and Cognition*, 43, 135–138.

De Brabander, B., and Boone, C. (1990). Sex differences in perceived locus of control. *Journal of Social Psychology*, 130, 271–272.

De Brabander, B., and Declerck, C. (2004). A possible role of central dopamine metabolism in individual differences in locus of control. *Journal of Personality and Individual Differences*, 37, 735–750.

De Brabander, B., Boone, C., and Gerits, P. (1992). Locus of control and cerebral asymmetry. *Perceptual and Motor Skills*, 75, 131–413.

Declerck, C. H., Boone, C., and De Brabander, B. (2006). On feeling in control: a biological theory for individual differences in control perception. *Brain and Cognition*, 62, 143–176.

Deglin, V. L., and Kinsbourne, M. (1996). Divergent thinking styles of the hemispheres: how syllogisms are solved during transitory hemisphere suppression. *Brain and Cognition*, 31, 285–307.

de la Fuente-Fernandez, R., Kishore, A., Calne, D. B., Ruth, T. J., and Stoessl, A. J. (2000). Nigrostriatal dopamine system and motor lateralization. *Behavioural and Brain Research*, 112, 63–68.

DeLange, F. (2000). The role of iodine in brain development. *Proceedings of the Nutrition Society*, 59, 75–79.

Delaveau, P., Salgado-Pineda, P., Wickerm B., Micallef-Roll, J., and Blin, O. (2005). Effect of levodopa on healthy volunteers' facial emotion perception: an FMRI study. *Clinical Neuropharmacology*, 28, 255–261.

Delion, S., Chalon, S., Guilloteau, D., Besnard, J. C., and Durand, G. (1996). Alpha-Linolenic acid dietary deficiency alters age-related changes of dopaminergic and serotoninergic neurotransmission in the rat frontal cortex. *Journal of Neurochemistry*, 66, 1582–1591.

Denk, F., Walton, M. E., Jennings, K. A., Sharp, T., Rushworth, M. F., and Bannerman, D. M. (2005). Differential involvement of serotonin and dopamine systems in cost-benefit decisions about delay or effort. *Psychopharmacology*, 179, 587–596.

Denys, D., Van Der Wee, N., Janssen, J., De Geus, F., and Westenberg, H. G. (2004). Low level of dopaminergic D2 receptor binding in obsessive-compulsive disorder. *Biological Psychiatry*, 55, 1041–1045.

Depue, R. A., and Collins, P. F. (1999). Neurobiology of the structure of personality: dopamine, facilitation of incentive motivation, and extraversion. *Behavioral and Brain Sciences*, 22, 491–569.

Depue, R. A., and Iacono, W. G. (1989). Neurobehavioral aspects of affective disorders. *Annual Review of Psychology*, 40, 457–492.

Depue, R. A., and Morrone-Strupinsky, J. V. (2005). A neurobehavioral model of affiliative bonding: implications for conceptualizing a human trait of affiliation. *Behavioral and Brain Sciences*, 28, 313–350.

Deutch, A. Y., Clark, W. A., and Roth, R. H. (1990). Prefrontal cortical dopamine depletion enhances the responsiveness of mesolimbic dopamine neurons to stress. *Brain Research*, 521, 311–315.

Diamond, A., Prevor, M. B., Callender, G., and Druin, D. P. (1997). Prefrontal cortex cognitive deficits in children treated early and continuously for PKU. *Monographs of the Society for Research in Child Development*, 62(4), 1–208.

Dickens, W. T., and Flynn, J. R. (2001). Heritability estimates versus large environmental effects: the IQ paradox resolved. *Psychological Review*, 108, 346–369.

Dinn, W. M., Harris, C. L., Aycicegi, A., Greene, P., and Andover, M. S. (2002). Positive and negative schizotypy in a student sample: neurocognitive and clinical correlates. *Schizophrenia Research*, 56, 171–185.

Dobson, J. E. (1998). The iodine factor in health and evolution. *The Geographical Review*, 88, 1–18.

Dodd, M. L., Klos, K. J., Bower, J. H., Geda, Y. E., Josephs, K. A., and Ahlskog, J. E. (2005). Pathological gambling caused by drugs used to treat Parkinson disease. *Archives of Neurology*, 62, 1377–1381.

Dorus, S., Vallender, E. J., Evans, P. D., Anderson, J. R., Gilbert, S. L., Mahowald, M., Wyckoff, G. J., Malcom, C. M., and Lahn, B. T. (2004). Accelerated evolution of nervous system genes in the origin of *Homo sapiens*. *Cell*, 119, 1027–1040.

Dringenberg, H. C., Wightman, M., and Beninger, R. J. (2000). The effects of amphetamine and raclopride on food transport: possible relation to defensive behavior in rats. *Behavioral Pharmacology*, 11, 447–554.

Dringenberg, H. C., Dennis, K. E., Tomaszek, S., and Martin, J. (2003). Orienting and defensive behaviors elicited by superior colliculus stimulation in rats: effects of 5-HT depletion, uptake inhibition, and direct midbrain or frontal cortex application. *Behavioural and Brain Research*, 144, 95–103.

Dulawa, S. C., Grandy, D. K., Low, M. J., Paulus, M. P., and Geyer, M. A (1999). Dopamine D4 receptor-knock-out mice exhibit reduced exploration of novel stimuli. *Journal of Neuroscience*, 19, 9550–9556.

Duncan, R. C. (2005). The Olduvai theory: energy, population, and industrial civilization. *The Social Contract*, 16(2), 134–144.

Easterbrook, G. (2004). Red scare. *The New Republic Online*, February 2, 2004. www.tnr.com/doc.mhtml?i=20040202&s=easterbrook020204.

Eaton, S. B. (1992). Humans, lipids and evolution. *Lipids*, 27, 814–820.

Eisen, J. L., and Rasmussen, S. A. (1993). Obsessive compulsive disorder with psychotic features. *Journal of Clinical Psychiatry*, 54, 373–379.

Elliott, F. A. (1982). Violence: a neurological overview. *Practitioner*, 226, 301–304.

Ellis, K. A., and Nathan, P. J. (2001). The pharmacology of human working memory. *International Journal of Neuropsychopharmacology*, 4, 299–313.

Enard, W., Przeworski, M., Fisher, S. E., Lai, C. S. L., Wiebe, V., Kitano, K., Monaco, A. P., and Paabo, S. (2002). Molecular evolution of FOXP2, a gene involved in speech and language. *Nature*, 418, 869–872.

Erlandson, J. M. (2001). The archaeology of aquatic adaptations: paradigms for a new millennium. *Journal of Archaeological Research*, 9, 287–350.

Evans, A. H., Pavese, N., Lawrence, A. D., Tai, Y. F., Appel, S., Doder, M., Brooks, D. J., Lees, A. J., and Piccini, P. (2006). Compulsive drug use linked to sensitized ventral striatal dopamine transmission. *Annals of Neurology*, 59, 852–858.

Fabisch, K., Fabisch, H., Langs, G., Huber, H. P., and Zapotoczky, H. G. (2001). Incidence of obsessive-compulsive phenomena in the course of acute schizophrenia and schizoaffective disorder. *European Psychiatry*, 16, 336–341.

Falcone, D. J., and Loder, K. (1984). A modified lateral eye-movement measure, the right hemisphere and creativity. *Perceptual and Motor Skills*, 58, 823–830.

Falk, D. (1990). Brain evolution in *Homo*: The "radiator" theory. *Behavioral and Brain Sciences*, 13, 333–344.

Fallon, J. H., and Loughlin, S. E. (1987). Monoamine innervation of cerebral cortex and a theory of the role of monoamines in cerebral cortex and basal ganglia. In E. G. Jones and A. Peters (eds), *Cerebral Cortex* (Vol. 6). New York: Plenum.

Faraone, S. V., Biederman, J., and Mick, E. (2006). The age-dependent decline of attention deficit hyperactivity disorder: a meta-analysis of follow-up studies. *Psychological Medicine*, 36, 159–165.

Farde, L., Gustavsson, J. P., and Jonsson, E. (1997). D2 dopamine receptors and personality traits. *Nature*, 383, 590.

Faridi, K., and Suchowersky, O. (2003). Gilles de la Tourette's Syndrome. *Canadian Journal of the Neurological Sciences*, 30 (Suppl. 1), S64–S71.

Farmer, M. E., and Klein, R. M. (1995). The evidence for a temporal processing deficit linked to dyslexia: a review. *Psychonomic Bulletin and Review*, 2, 460–493.

Fellows, L. K., and Farah, M. J. (2005). Dissociable elements of human foresight: a role for the ventromedial frontal lobes in framing the future, but not in discounting future rewards. *Neuropsychologia*, 43, 1214–1221.

Fenwick, P., Galliano, S., Coate, M. A., Rippere, V., and Brown, D. (1985). "Psychic" sensitivity, mystical experience, head injury and brain pathology. *British Journal of Medical Psychology*, 58, 35–44.

Fibiger, H. C., Phillips, A. G., and Brown, E. E. (1992). The neurobiology of cocaine-induced reinforcement *Ciba Foundation Symposium*, 166, 96–111.

Figa-Talamanca, L., and Gualandi, C. (1989). Hyperthermic syndromes and impairment of the dopaminergic system: a clinical study. *Italian Journal of Neurological Sciences*, 10, 49–59.

Fink, J. S., and Smith, G. P. (1980). Mesolimbic and mesocortical dopaminergic neurons are necessary for normal exploratory behavior in rats. *Neuroscience Letters*, 17, 61–65.

Finlay, J. M., and Zigmond, M. J. (1997). The effects of stress on central dopaminergic neurons: possible clinical implications. *Neurochemical Research*, 22, 1387–1394.

Fitch, W. T. (2006). The biology and evolution of music: a comparative perspective. *Cognition*, 100, 173–215.

Flaherty, A. W. (2005). Frontotemporal and dopaminergic control of idea generation and creative drive. *Journal of Comparative Neurology*, 493, 147–153.

Flor-Henry, P. (1986). Observations, reflections and speculations on the cerebral determinants of mood and on the bilaterally asymmetrical distributions of the major neurotransmitter systems. *Acta Neurologica Scandinavica*, 74 (Suppl. 109), 75–89.

Folk, G. E., Jr., and Semken, H. A., Jr. (1991). The evolution of sweat glands. *International Journal of Biometeorology*, 35, 180–186.

Freeman, M. P., Freeman, S. A., and Mcelroy, S. L. (2002). The comorbidity of bipolar and anxiety disorders: prevalence, psychobiology, and treatment issues. *Journal of Affective Disorders*, 68, 1–23.

Freud, S. (1927). *The Ego and the Id* (trans. J. Riviere). London: Hogarth Press and Institute of Psycho-Analysis.

Fry, A. F., and Hale, S. (2000). Relationships among processing speed, working memory, and fluid intelligence in children. *Biological Psychology*, 54, 1–34.

Fukumura, M., Cappon, G. D., Broening, H. W., and Vorhees, C. V. (1998). Methamphetamine-induced dopamine and serotonin reductions in neostriatum are not gender specific in rats with comparable hyperthermic responses. *Neurotoxicology and Teratology*, 20, 441–448.

Gagneux, P., Arness, B., Diaz, S., Moore, S., Patel, T., Dillmann, W., Parkeh, R., and Varki, A. (2001). Proteomic comparison of human and great ape blood plasma reveals conserved glycosylation and differences in thyroid hormone metabolism. *American Journal of Physical Anthropology*, 115, 99–109.

Galvan, A., Hare, T., Voss, H., Glover, G., and Casey, B. J. (2007). Risk-taking and the adolescent brain: who is at risk? *Developmental Science*, 10, F8–F14.

Garner, J. P., Meehan, C. L., and Mench, J. A. (2003). Stereotypies in caged parrots, schizophrenia and autism: evidence for a common mechanism. *Behavioural Brain Research*, 145, 125–134.

Gaspar, P., Berger, B., Febvret, A., Vigny, A., and Henry, J. P. (1989). Catecholamine innervation of the human cerebral cortex as revealed by comparative immunohistochemistry of tyrosine hydroxylase and dopamine-beta-hydroxylase. *Journal of Comparative Neurology*, 279, 249–271.

Gazzaniga, M. S. (1983). Right hemisphere language following brain bisection: a 20-year perspective. *American Psychologist*, 38, 525–537.

Gazzaniga, M. S. (2005). Forty-five years of split-brain research and still going strong. *Nature Review Neuroscience*, 6, 653–659.

Geller, D. A., Biederman, J., Faraone, S., Spencer, T., Doyle, R., Mullin, B., Magovcevic, M., Zaman, N., and Farrell, C. (2004). Re-examining comorbidity of obsessive compulsive and attention-deficit hyperactivity disorder using an empirically derived taxonomy. *European Child and Adolescent Psychiatry*, 13, 83–91.

Gervais, J., and Rouillard, C. (2000). Dorsal raphe stimulation differentially modulates dopaminergic neurons in the ventral tegmental area and substantia nigra. *Synapse*, 35, 281–291.

Geschwind, N., and Galaburda, A. M. (1985). Cerebral lateralization. Biological mechanisms, associations, and pathology. I. A hypothesis and program for research. *Archives of Neurology*, 42, 428–459.

Gilbert, C. (1995). Optimal physical performance in athletes: key roles of dopamine in a specific neurotransmitter/hormonal mechanism. *Mechanisms of Ageing and Development*, 84, 83–102.

Gillam, M. P., Fideleff, H., Boquete, H. R., and Molitch, M. E. (2004). Prolactin excess: treatment and toxicity. *Pediatric Endocrinology Review*, 2 (Suppl. 1), 108–114.

Gillberg, C., and Billstedt, E. (2000). Autism and Asperger syndrome: coexistence with other clinical disorders. *Acta Psychiatrica Scandinavica*, 102, 321–339.

Girard, T. A., and Cheyne, J. A. (2004). Individual differences in lateralisation of hallucinations associated with sleep paralysis. *Laterality*, 9, 93–111.

Girard, T. A., Martius, D. L. M. A., and Cheyne, J. A. (2007). Mental representation of space: insights from an oblique distribution of hallucinations. *Neuropsychologia*, 45, 1257–1269.

Giuliano, F., and Allard, J. (2001). Dopamine and male sexual function. *European Urology*, 40, 601–608.

Gleick, J. (2003). *Isaac Newton*. New York: Pantheon Books.

Glickstein, M., and Stein, J. (1991). Paradoxical movement in Parkinson's disease. *Trends in Neurosciences*, 14, 480–482.

Goldman-Rakic, P. S. (1998). The cortical dopamine system: role in memory and cognition. *Advances in Pharmacology*, 42, 707–711.

Goldstein, R. Z., and Volkow, N. D. (2002). Drug addiction and its underlying neurobiological basis: neuroimaging evidence for the involvement of the frontal cortex. *American Journal of Psychiatry*, 159, 1642–1652.

Goodman, N. (2002). The serotonergic system and mysticism: could LSD and the nondrug-induced mystical experience share common neural mechanisms? *Journal of Psychoactive Drugs*, 34, 263–272.

Goodwin, F. K., and Jamison, K. R. (1990). *Manic-Depressive Illness*. New York: Oxford University Press.

Gopnik, M. (1990). Genetic basis of grammar defect. *Nature*, 347, 26.

Gottesmann C. (2002). The neurochemistry of waking and sleeping mental activity: the disinhibition–dopamine hypothesis. *Psychiatry and Clinical Neurosciences*, 56, 345–354.

Gottlieb, G. (1998). Normally occurring environmental and behavioral influences on gene activity: from central dogma to probabilistic epigenesis. *Psychological Review*, 105, 792–802.

Gottlieb, K., and Manchester, D. K. (1986). Twin study methodology and variability in xenobiotic placental metabolism. *Teratogenesis, Carcinogenesis, and Mutagenesis*, 6, 253–263.

Gray, J. (1992). *Men are from Mars, Women are from Venus: the Classic Guide to Understanding the Opposite Sex*. New York: Harper Collins.

Gray, J. R., Chabris, C. F. and Braver, T. S. (2003). Neural mechanisms of general fluid intelligence. *Nature Neuroscience*, 6, 316–322.

Greene, A. L., and Wheatley, S. M. (1992). "I've got a lot to do and I don't think I'll have the time": Gender differences in late adolescents' narratives of the future. *Journal of Youth and Adolescence*, 21, 667–686.

Greenfield, P. M. (1991). Language, tools and brain: the ontogeny and phylogeny of hierarchically organized sequential behavior. *Behavioral and Brain Sciences*, 14, 531–551.

GreenFieldboyce, N. (2006). Study: some key materials growing scarce (www.npr.org/templates/story/story.php?storyId=5159755).

Grine, F. E., and Kay, R. F. (1988). Early hominid diets from quantitative image analysis of dental microwear. *Nature*, 333, 765–768.

Gross-Isseroff, R., Hermesh, H., and Weizman, A. (2001). Obsessive compulsive behaviour in autism – towards an autistic-obsessive compulsive syndrome? *World Journal of Biological Psychiatry*, 2, 193–197.

Gruzelier, J. H. (1999). Functional neuropsychological asymmetry in schizophrenia: a review and reorientation. *Schizophrenia Bulletin*, 25, 91–120.

Gunter, P. R. (1983). Religious dreaming: a viewpoint. *American Journal of Psychiatry*, 37, 411–427.

Gunturkun, O. (2005). The avian 'prefrontal cortex' and cognition. *Current Opinion in Neurobiology*, 15, 686–693.

Guo, J. F., Yang, Y. K., Chiu, N. T., Yeh, T. L., Chen, P. S., Lee, I. H., and Chu, C. L. (2006). The correlation between striatal dopamine D2/D3 receptor availability and verbal intelligence quotient in healthy volunteers. *Psychological Medicine*, 36, 547–554.

Gur, R. E., and Chin, S. (1999). Laterality in functional imaging studies of schizophrenia. *Schizophrenia Bulletin*, 25, 141–156.

Hall, A., and Walton, G. (2004). Information overload within the health care system: a literature review. *Health Information and Libraries Journal*, 21, 102–108.

Hall, S., and Schallert, T. (1988). Striatal dopamine and the interface between orienting and ingestive functions. *Physiology and Behavior*, 44, 469–471.

Hamblin, D. J. (1987). Has the Garden of Eden been located at last? *Smithsonian*, 18, 127–135.

Hammer, M. F. (1995). A recent common ancestry for human Y chromosomes. *Nature*, 378, 376–378.

Hammond, N. G. L. (1997). *The Genius of Alexander the Great*. Chapel Hill: University of North Carolina Press.

Handwerk, B. (2006). "Python cave" reveals oldest human ritual, scientists suggest. *National Geographic News*, December 22, 2006 (http://news. nationalgeographic.com/news/2006/12/061222-python-ritual.html).

Hanley, W. B., Koch, R., Levy, H. L., Matalon, R., Rouse, B., Azen, C., and de la Cruz, F. (1996). The North American maternal phenylketonuria collaborative study, developmental assessment of the offspring: preliminary report. *European Journal of Pediatrics*, 155 (Suppl. 1), S169–S172.

Happe, F., Brownwell, H., and Winner, E. (1999). Acquired "theory of mind" impairments following stroke. *Cognition*, 70, 211–240.

Hardyck, C., Petrinovich, L. F., and Goldman, R. D. (1976). Left-handedness and cognitive deficit. *Cortex*, 12, 266–279.

Hare, E. H., and Walter, S. D. (1978). Seasonal variation in admissions of psychiatric patients and its relation to seasonal variation in their births. *Journal of Epidemiology and Community Health*, 32, 47–52.

Harper, L. V. (2005). Epigenetic inheritance and the intergenerational transfer of experience. *Psychological Bulletin*, 131, 340–360.

Hasan, W., Cowen, T., Barnett, P. S., Elliot, E., Coskeran, P., and Bouloux, P. M. (2001). The sweating apparatus in growth hormone deficiency, following treatment with r-hGH and in acromegaly. *Autonomic Neuroscience*, 89, 100–109.

Hawkes, K., and O'Connell, J. F. (1981) Affluent hunters? Some comments in light of the Alyawara case. *American Anthropologist*, 83, 622–626.

Hawks, J., Hunley, K., Lee, S. H., and Wolpoff, M. (2000). Population bottlenecks and Pleistocene human evolution. *Molecular Biology and Evolution*, 17, 2–22.

Hayashi, M., Kato, M., Igarashi, K., and Kashima, H. (2008). Superior fluid intelligence in children with Asperger's disorder. *Brain and Cognition*, 66, 306–310.

Hecaen, H., and Albert, M. L. (1978). *Human Neuropsychology*. New York: Wiley & Sons.

Heilman, K. M., and Gilmore, R. L. (1998). Cortical influences in emotion. *Journal of Clinical Neurophysiology*, 15, 409–423.

Heilman, K. M., Voeller, K. K., and Nadeau, S. E. (1991). A possible pathophysiologic substrate of attention deficit hyperactivity disorder. *Journal of Child Neurology*, 6 (Suppl.), S76–S81.

Hein, A. (1974). Prerequisite for development of visually guided reaching in the kitten. *Brain Research*, 71, 259–263.

Henneberg, M. (1998). Evolution of the human brain: is bigger better? *Clinical Experimental Pharmacology and Physiology*, 25, 745–749.

Henshilwood, C. S., d'Errico, F., Marean, C. W., Milo, R. G., and Yates, R. (2001). An early bone tool industry from the Middle Stone Age at Blombos Cave, South Africa: implications for the origins of modern human behaviour, symbolism and language. *Journal of Human Evolution*, 41, 631–678.

Herman, L. M. (1986). Cognition and language competencies of bottlenosed dolphins. In R. J. Schusterman, J. A. Thomas and F. G. Wood (eds), *Dolphin Cognition and Behavior: a Comparative Approach* (pp. 221–252). Hillsdale: Erlbaum.

Hershman, J., Hershman, D. J., and Lieb, J. (1994). *A brotherhood of tyrants*. New York: Prometheus Books.

Heyes, M. P., Garnett, E. S., and Coates, G. (1988). Nigrostriatal dopaminergic activity is increased during exhaustive exercise stress in rats. *Life Sciences*, 42, 1537–1542.

Hills, T. T. (2006). Animal foraging and the evolution of goal-directed cognition. *Cognitive Science*, 30, 3–41.

Hills, T. T., Todd, P. M., and Goldstone, R. L. (2007). Implications for human cognition from the evolution of animal foraging. *Proceedings of the 19th Annual Meeting of the Human Behavior and Evolution Society*, Williamsburg, VA.

Hobson, J. A. (1996). How the brain goes out of its mind. *Endeavour*, 20, 86–89.

Hobson, J. A., Pace-Schott, E. F., and Stickgold, R. (2000). Dreaming and the brain: toward a cognitive neuroscience of conscious states. *Behavioral and Brain Sciences*, 23, 793–1121.

Hockett, C. F. (1960). Logical considerations in the study of animal communication. In W. E. Lanyon and W. N. Tavolga (eds), *Animal Sounds and Communication* (pp. 392–430). Washington, DC: American Institute of Biological Sciences.

Holloway, R. L. (2002). Brief communication: how much larger is the relative volume of Area 10 of the prefrontal cortex in humans? *American Journal of Physical Anthropology*, 118, 399–401.

Holmberg, T. (2002). *FAQs: Napoleon and the French Revolution*. In the Napoleon Series (www.napoleonseries.org).

Hoppe, K. D. (1988). Hemispheric specialization and creativity. *Psychiatric Clinics of North America*, 11, 303–315.

Horgan, J. (1996). *The End of Science*. New York: Broadway Books.

Horner, A. J. (2006). The unconscious and the creative process. *Journal of the American Academy of Psychoanalysis and Dynamic Psychiatry*, 34, 461–469.

Horrobin, D. F. (1998). Schizophrenia: the illness that made us human. *Medical Hypotheses*, 50, 259–288.

Horvitz, J. C. (2000). Mesolimbic and nigrostriatal dopamine responses to salient non-reward events. *Neuroscience*, 96, 651–656.

Hotson, J. R., Langston, E. B., and Langston, J. W. (1986). Saccade responses to dopamine in human MPTP-induced Parkinsonism. *Annals of Neurology*, 20, 456–463.

Howard, G. S. (2000). Adapting human lifestyles for the 21st century. *American Psychologist*, 55, 509–515.

Hull, E., Muschamp, J. W., and Sato, S. (2004). Dopamine and serotonin: influences on male sexual behavior. *Physiology and Behavior*, 83, 291–307.

Huseman, C. A., Hassing, J. M., and Sibilia, M. G. (1986). Endogenous dopaminergic dysfunction: a novel form of human growth hormone deficiency and short stature. *Journal of Clinical Endocrinology and Metabolism*, 62, 484–490.

Ikemoto, S. (2007). Dopamine reward circuitry: two projection systems from the ventral midbrain to the nucleus accumbens – olfactory tubercle complex. *Brain Research Reviews*, 56, 27–78.

Ikemoto, S., and Panksepp, J. (1999). The role of nucleus accumbens dopamine in motivated behavior: a unifying interpretation with special reference to reward-seeking. *Brain Research Reviews*, 31, 6–41.

Ipsen, D. C. (1985). *Isaac Newton: Reluctant Genius*. Hillside: Enslow Publishers.

Isaacson, W. (2007). *Einstein: His Life and Universe*. Simon and Schuster.

Iversen, S. D. (1984). Cortical monoamines and behavior. In L. Descarries, T. R. Reader and H. H. Jasper (eds), *Monoamine Innervation of Cerebral Cortex* (pp. 321–351). New York: Liss.

Jablonski, N. G. (2004). The evolution of human skin and skin color. *Annual Review of Anthropology*, 33, 585–623.

Jacoby, J. H., Greenstein, M., Sassin, J. F., and Weitzman, E. D. (1974). The effect of monoamine precursors on the release of growth hormone in the rhesus monkey. *Neuroendocrinology*, 14, 95–102.

James, H. V. A., and Petraglia, M. D. (2005). Modern human origins and the evolution of behavior in the later Pleistocene record of South Asia. *Current Anthropology*, 46, S3–S16.

Jancke, L., and Steinmetz, H. (1994). Auditory lateralization in monozygotic twins. *International Journal of Neuroscience*, 75, 57–64.

Jankovic, J. (2002). Levodopa strengths and weaknesses. *Neurology*, 58, S19–S32.

Jaspers, K. (1964). *General Psychopathology* (trans. J. Hoenig and M. W. Hamilton). Chicago: University of Chicago Press.

Jaynes, J. (1976). *The Origins of Consciousness in the Breakdown of the Bicameral Mind*. Boston: Houghton Mifflin.

Johnson, B. A. (2004). Role of the serotonergic system in the neurobiology of alcoholism: implications for treatment. *CNS Drugs*, 18, 1105–1118.

Johnson, F. W. (1991). Biological factors and psychometric intelligence: a review. *Genetic, Social and General Psychology Monographs*, 117, 313–357.

Johnson, P. (2002). *Napoleon*. New York: Viking Penguin.

Joseph, R. (1999). Frontal lobe psychopathology: mania, depression, confabulation, catatonia, perseveration, obsessive compulsions, and schizophrenia. *Psychiatry*, 62, 138–172.

Kalbag, A. S., and Levin, F. R. (2005). Adult ADHD and substance abuse: diagnostic and treatment issues. *Substance Use and Misuse*, 40, 1955–1981, 2043–2048.

Kalsbeek, A., De Bruin, J. P., Feenstra, M. G., Matthijssen, M. A., and Uylings, H. B. (1988). Neonatal thermal lesions of the mesolimbocortical dopaminergic projection decrease food-hoarding behavior. *Brain Research*, 475, 80–90.

Kanazawa, S. (2003). Why productivity fades with age: the crime-genius connection. *Journal of Research in Personality*, 37, 257–272.

Kane, M. J., and Engle, R. W. (2002). The role of prefrontal cortex in working-memory capacity, executive attention, and general fluid intelligence: an individual differences perspective. *Psychonomic Bulletin and Review*, 9, 637–671.

Kapur, S. (2003). Psychosis as a state of aberrant salience: a framework linking biology, phenomenology, and pharmacology in schizophrenia. *American Journal of Psychiatry*, 160, 13–23.

Karlsson, J. L. (1974). Inheritance of schizophrenia. *Acta Psychiatrica Scandinavica* (Suppl. 247), 1–116.

Kaulins, A. (1979). Cycles in the birth of eminent humans. *Cycles*, 30, 9–15.

Kay, R. F. Cartmill, M., and Balow, M. (1998). The hypoglossal canal and the origin of human vocal behavior. *Proceedings of the National Academy of Sciences*, 95, 5417–5419.

Keenan, J. P., Wheeler, M. A., Gallup, G. G., Jr., and Pascual-Leone, A. (2000). Self-recognition and the right prefrontal cortex. *Trends in Cognitive Sciences*, 4, 338–344.

Keller, E. F. (2000). *The Century of the Gene*. Cambridge, MA: Harvard University Press.

Kelso, J. A. S., and Tuller, B. (1984). Converging evidence in support of common dynamical principles for speech and movement coordination. *American Journal of Physiology*, 246, R928–R935.

Kenealy, P. M. (1996). Pheylketonuria. In J. G. Beaumont, P. M. Kenealy, and M. J. C. Rogers (eds), *The Blackwell Dictionary of Neuropsychology* (pp. 570–575). Cambridge, MA: Blackwell Publishers.

Kennedy, N., Boydell, J., Kalidindi, S., Fearon, P., Jones, P. B., Van Os, J., and Murray, R. M. (2005). Gender differences in incidence and age at onset of mania and bipolar disorder over a 35-year period in Camberwell, England. *American Journal of Psychiatry*, 162, 257–262.

Kerbeshian, J., Burd, L., and Klug, M. G. (1995). Comorbid Tourette's disorder and bipolar disorder: an etiologic perspective. *American Journal of Psychiatry*, 152, 1646–1651.

Kimura, D. (1993). *Neuromotor Mechanisms in Human Communication.* New York: Oxford University Press.

Kingston, J. (1990). Five exaptations in speech: reducing the arbitrariness of the constraints on language. *Behavioral and Brain Sciences*, 13, 738–739.

Kiyohara, T., Hori, T., Shibata, M., Kakashima, T., and Osaka, T. (1984). Neuronal inputs to preoptic thermosensitive neurons – histological and electrophysiological mapping of central connections. *Journal of Thermal Biology*, 9, 21–26.

Klawans, H. L. (1987). Chorea. *Canadian Journal of the Neurological Sciences*, 14, 536–540.

Kluger, A. N., Siegfried, Z., and Ebstein, R. P. (2002). A meta-analysis of the association between DRD4 polymorphism and novelty seeking. *Molecular Psychiatry*, 7, 712–717.

Koepp, M. J., Gunn, R. N., Lawrence, A. D., Cunningham, V. J., Dagher, A., Jones, T., Brooks, D. J., Bench, C. J., and Grasby, P. M. (1998). Evidence for striatal dopamine release during a video game. *Nature*, 393, 266–268.

Koob, G. F., Riley, S. J., Smith, S. C., and Robbins, T. W. (1978). Effects of 6-hydroxydopamine lesions of the nucleus acumbens septi and olfactory tubercle on feeding, locomotor activity, and amphetamine anorexia in the rat. *Journal of Comparative and Physiological Psychology*, 92, 917–927.

Koyama, S. (1995). Japanese dreams: culture and cosmology. *Psychiatry and Clinical Neurosciences*, 49, 99–101.

Krantz, G. S. (1968). Brain size and hunting ability in earliest man. *Current Anthropology*, 9, 450–451.

Krause, J., Lalueza-Fox, C., Orlando, L., Enard, W., Green, R. E., Burbano, H. A., Hublin, J.-J., Hanni, C., Fortea, J., de la Rasilla, M., Bertranpetit, J., Rosas, A., and Paabo, S. (2007). The derived FoxP2 variant of modern humans was shared with Neandertals. *Current Biology Online*, October 18, 2007.

Krystal, J. H., Perry, E. B., Jr., Gueorguieva, R., Belger, A., Madonick, S. H., Abi-Dargham, A., Cooper, T. B. L., Macdougall, W., Abi-Saab, and D'Souza, D. C. (2005). Comparative and interactive human psycho-pharmacologic effects of ketamine and amphetamine: implications for glutamatergic and dopaminergic model psychoses and cognitive function. *Archives of General Psychiatry*, 62, 985–994.

Kucharska-Pietura, K., Phillips, M. L., Gernand, W., and David, A. S. (2003). Perception of emotions from faces and voices following unilateral brain damage. *Neuropsychologia*, 41, 1082–1090

Kushner, M. G., Riggs, D. S., Foa, E. B., and Miller, S. M. (1993). Perceived controllability and the development of posttraumatic stress disorder (PTSD) in crime victims. *Behavioral Research and Therapy*, 31, 105–110.

Kyllonen, P. C., and Christal, R. E. (1990). Reasoning ability is (little more than) working-memory capacity?! *Intelligence*, 14, 389–433.

Lai, C. S., Fisher, S. E., Hurst, J. A., Vargha-Khadem, F., and Monaco, A. P. (2001). A forkhead-domain gene is mutated in a severe speech and language disorder. *Nature*, 413, 519–523.

Laland, K. N., Odling-Smee, F. J., and Feldman, M. W. (1999). Evolutionary consequences of niche construction and their implications for ecology. *Proceedings of the Natural Academy of Sciences*, 96, 10242–10247.

Lange, M., Thulesen, L., Feldt-Rasmussen, U., Skakkebaek, N. E., Vahl, N., Jorgensen, J. O., Christiansen, J. S., Poulsen, S. S., Sneppen, S. B., and Juul, A. (2001). Skin morphological changes in growth hormone deficiency and acromegaly. *European Journal of Endocrinology*, 145, 147–153.

Larisch, R., Meyer, W., Klimke, A., Kehren, F., Vosberg, H., and Muller-Gartner, H. W. (1998). Left-right asymmetry of striatal dopamine D2 receptors. *Nuclear Medicine Communications*, 19, 781–787.

Lauritsen, M. B., Pedersen, C. B., and Mortensen, P. B. (2005). Effects of familial risk factors and place of birth on the risk of autism: a nationwide register-based study. *Journal of Child Psychology and Psychiatry*, 46, 963–971.

Leakey, R., and Lewin, R. (1995). *The Sixth Extinction: Biodiversity and its Survival*. New York: Doubleday.

Lee, A. C., Harris, J. P., Atkinson, E. A., Nithi, K., and Fowler, M. S. (2002). Dopamine and the representation of the upper visual field: evidence from vertical bisection errors in unilateral Parkinson's disease. *Neuropsychologia*, 40, 2023–2029.

Lee, R. B. (1979). *The !Kung San: Men, Women, and Work in a Foraging Society*. Cambridge: Cambridge University Press.

Lee, T. F., Mora, F., and Myers, R. D. (1985). Dopamine and thermoregulation: an evaluation with special reference to dopaminergic pathways. *Neuroscience and Biobehavioral Reviews*, 9, 589–598.

Leonard, W. R., and Robertson, M. L. (1997). Comparative primate energetics and hominid evolution. *American Journal of Physical Anthropology*, 102, 265–281.

Lewis D. A., and Levitt, P. (2002). Schizophrenia as a disorder of neurodevelopment. *Annual Review of Neuroscience*, 25, 409–432.

Li, D., Sham, P. C., Owen, M. J., and He, L. (2006). Meta-analysis shows significant association between dopamine system genes and attention deficit hyperactivity disorder (ADHD). *Human Molecular Genetics*, 15, 2276–2284.

Lickliter, R, and Honeycutt, H. (2003). Developmental dynamics: toward a biologically plausible evolutionary psychology. *Psychological Bulletin*, 129, 819–835.

Lieberman, P., Kako, E., Friedman, J., Tajchman, G., Feldman, L. S., and Jiminez, E. B. (1992). Speech production, syntax comprehension, and cognitive deficits in Parkinson's disease. *Brain and Language*, 43, 169–189.

Lightman, A (2004). Einstein and Newton: genius compared. *Scientific American*, 290(9), 108–109.

Litvan, I. (1996). Parkinson's disease. In J. G. Beaumont, P. M. Kenealy, and M. J. C. Rogers (eds), *The Blackwell Dictionary of Neuropsychology* (pp. 559–564). Cambridge, MA: Blackwell Publishers.

Lock, A., and Colombo, M. (1996). Cognitive abilities in a comparative perspective. In A. Lock and C. Peters (eds), *Handbook of Human Symbolic Evolution* (pp. 595–643). Oxford: Clarendon.

Lombroso, P. J., Mack, G., Scahill, L., King, R. A., and Leckman, J. F. (1991). Exacerbation of Gilles de la Tourette's syndrome associated with thermal stress: a family study. *Neurology*, 41, 1984–1987.

Lorber, J. (1983). Is your brain really necessary? In D. Voth (ed.), *Hydrocephalus im fruhen Kindesalter: Fortschritte der Grundlagenforschung, Diagnostik und Therapie* (pp. 2–14). Stuttgart: Enke.

Luciana, M., Collins, P. F., and Depue, R. A. (1998). Opposing roles for dopamine and serotonin in the modulation of human memory. *Cerebral Cortex*, 8, 218–226.

Macaulay, V., Hill, C., Achilli, A., Rengo, C., Clark, D., Meehan, W., *et al.* (2005). Single, rapid coastal settlement of Asia revealed by analysis of complete mitochondrial genomes. *Science*, 308, 1034–1036.

McBrearty, S., and Brooks, A. S. (2000). The revolution that wasn't: a new interpretation of the origin of modern human behavior. *Journal of Human Evolution*, 39, 453–563.

McCarthy, S. (2003). Water scarcity could affect billions: is this the biggest crisis of all? *Independent/UK*, March 5, 2003 (www.commondreams.org/headlines03/0305–05.htm).

McClure, S. M., Laibson, D. I., Loewenstein, G., and Cohen, J. D. (2004). Separate neural systems value immediate and delayed monetary rewards. *Science*, 306, 503–507.

MacDonald, D. A., and Holland, D. (2002). Spirituality and complex partial epileptic–like signs. *Psychological Reports*, 91, 785–792.

McDougall, I., Brown, F. K., and Fleagle, J. G., (2005). Stratigraphic placement and age of modern humans from Kibish, Ethiopia. *Nature*, 433, 733–736.

McElroy, S. L., Phillips, K. A., and Keck, P. E., Jr. (1994). Obsessive compulsive spectrum disorder. *Journal of Clinical Psychiatry* (Suppl.), 33–51.

McGuigan, F. J. (1966). *Thinking: Studies of Covert Language Processes*. New York: Appleton-Century-Crofts.

McKelvey, J. R., Lambert, R., Mottron, L., and Shevell, M. I. (1995). Right-hemisphere dysfunction in Asperger's syndrome. *Journal of Child Neurology*, 10, 310–314.

McNamara, K. J. (ed.) (1995). *Evolutionary Change and Heterochrony*. New York: Wiley.

Maia, D. P. Teixeira, A. L., Jr., Quintao Cunningham, M. C., and Cardoso, F. (2005). Obsessive compulsive behavior, hyperactivity, and attention deficit disorder in Sydenham chorea. *Neurology*, 64, 1799–1801.

Mandel, J.-L. (1996). The human genome. In J.-P. Changeux and J. Chavaillon (eds), *Origins of the Human Brain* (pp. 94–126). Oxford: Clarendon.

Mandell, A. J. (1980). Toward a psychobiology of transcendance: God in the brain. In J. M. Davidson and R. J. Davidson (eds), *The Psychobiology of Consciousness* (pp. 379–464). New York: Plenum.

Mandler, G. (2001). Apart from genetics: what makes monozygotic twins similar? *The Journal of Mind and Behavior*, 22, 147–160.

Marcellis, M., Takei, N., and Van Os, J. (1999). Urbanization and risk for schizophrenia: does the effect operate before or around the time of illness onset? *Psychological Medicine*, 29, 1197–1203.

Marean, C. W., Bar-Matthews, M., Bernatchez, J., Fisher, E., Goldberg, P., Herries, A. I., Jacobs, Z., Jerardino, A., Karkanas, P., Minichillo, T., Nilssen, P. J., Thompson, E., Watts, I., and Williams, H. M. (2007). Early human use of marine resources and pigment in South Africa during the Middle Pleistocene. *Nature*, 449, 905–908.

Markus, H., and Nurius, P. (1986). Possible selves. *American Psychologist*, 41, 954–969.

Meck, W. H. (1996). Neuropharmacology of timing and time perception. *Cognitive Brain Research*, 3, 227–242.

Mecoy, L. (2002). Nature waits on fate of dam. *Sacramento Bee*, November 19, 2002 (www.arroyoseco.org/SB021119.htm).

Mega, M. S., and Cummings, J. L. (1994). Frontal-subcortical circuits and neuropsychiatric disorders. *Journal of Neuropsychiatry and Clinical Neuroscience*, 6, 358–370.

Meier, B. P., and Robinson, M. D. (2004). Why the sunny side is up: associations between affect and vertical position. *Psychological Science*, 15, 243–247.

Meier, B. P. Hauser, D. J., Robinson, M. D., Friesen, C. K., and Schjeldahl, K. (2007). What's "up" with god? Vertical space as a representation of the divine. *Journal of Personality and Social Psychology*, 93, 699–710.

Mellars, P. (2006). Why did modern human populations disperse from Africa ca. 60,000 years ago? A new model. *Proceedings of the National Academy of Sciences*, 103, 9381–9386

Melnick, M., Myrianthopoulos, N. C., and Christian, J. C. (1978). The effects of chorion type on variation in IQ in the NCPP twin population. *American Journal of Human Genetics*, 30, 425–433.

Miczek, K. A., Fish, E. W., de Bold, J. F., and de Almeida, R. M. M. (2002). Social and neural determinants of aggressive behavior: pharmacotherapeutic targets at serotonin, dopamine and γ-aminobutyric acid systems. *Psychopharmacology*, 163, 434–458.

Miller, M. T., Stromland, K., Ventura, L., Johansson, M., Bandim, J. M., and Gillberg, C. (2005). Autism associated with conditions characterized by developmental errors in early embryogenesis: a mini review. *International Journal of Developmental Neuroscience*, 23, 201–219.

Mintz, M., and Myslobodsky, M. S. (1983). Two types of hemisphere imbalance in hemi-Parkinsonism coded by brain electrical activity and electrodermal activity. In M. S. Myslobodsky (ed.), *Hemisyndromes: Psychology, Neurology, Psychiatry* (pp. 213–238). San Diego: Academic Press.

Mithen, S. (1996). *The Prehistory of the Mind: the Cognitive Origins of Art and Science*. London: Thames and Hudson.

Moeller, F. G., Barratt, E. S., Dougherty D. M., Schmitz, J. M., and Swann, A. C. (2001). Psychiatric aspects of impulsivity. *American Journal of Psychiatry*, 158, 1783–1793.

Moldan, B., Hak, T., Kovanda, J., Havranek, M., and KusKova, P. (2004). Composite indicators of environmental sustainability. Paper presented at *Statistics, Knowledge and Policy OECD World Forum on Key Indicators*. Palermo, November 10–13, 2004 (www.oecd.org/dataoecd/43/48/33829383. doc+Moldan+2004+GDP+ecological+footprint&hl=en&gl=us&ct=clnk& cd=1).

Moll, J., de Oliveira-Souza, R., Moll, F. T., Bramati, I. E., and Andreiuolo, P. A. (2002). The cerebral correlates of set-shifting: an fMRI study of the trail making test. *Arquivos de Neuro-Psiquiatria*, 60, 900–905.

Monchi, O., Ko, J. H., and Strafella, A. P. (2006). Striatal dopamine release during performance of executive functions: A[11C] raclopride PET study. *Neuroimage*, 33, 907–912.

Moncrieff, J., Wessely, S., and Hardy, R. (2004). Active placebos versus antidepressants for depression. *Cochrane Database System Review*, (1): CD003012.

Moore, J. The aquatic ape theory: sink or swim? (www.aquaticape.org).

Morgan, E. (1997). *The Aquatic Ape Hypothesis*. London: Souvenir Press.

Morison, S. E. (1983). *Admiral of the Ocean Sea: a Life of Christopher Columbus*. Boston: Northeastern University Press.

Morneau, D. M., Macdonald, D. A., and Holland, C. J. (1996). A confirmatory study of the relation between self-reported complex partial epileptic signs, peak experiences and paranormal beliefs. *British Journal of Clinical Psychology*, 35, 627–630.

Mozley, L. H., Gur, R. C., Mozley, P. D., and Gur, R. E. (2001). Striatal dopamine transporters and cognitive functioning in healthy men and women. *American Journal of Psychiatry*, 158, 1492–1499.

Muir, H. (2003). Did Einstein and Newton have autism? *New Scientist*, 2393, 10.

Muldoon, M. F., Mackey, R. H., Korytkowski, M. T., Flory, J. D., Pollock, B. G., and Manuck, S. B. (2006). The metabolic syndrome is associated with reduced central serotonergic responsivity in healthy community volunteers. *Journal of Clinical Endocrinology and Metabolism*, 91, 718–721.

Muller, N., Riedel, M., Zawta, P., Gunther, W., and Straube, A. (2002). Comorbidity of Tourette's syndrome and schizophrenia – biological and physiological parallels. *Progress in Neuropsychopharmacology and Biological Psychiatry*, 26, 1245–1252.

Murphy, M. B. (2000). Dopamine: a role in the pathogenesis and treatment of hypertension. *Journal of Human Hypertension*, 14, S47–S50.

Murrell, A., and Mingrone, M. (1994). Correlates of temporal perspective. *Perceptual and Motor Skills*, 78, 1331–1334.

Muzio, J. N., Roffwarg, H. P., and Kaufman, E. (1966). Alterations in the nocturnal sleep cycle resulting from LSD. *Electroencephalography and Clinical Neurophysiology*, 21, 313–324.

Mychack, P., Kramer, J. H., Boone, K. B., and Miller, B. L. (2001). The influence of right frontotemporal dysfunction on social behavior in frontotemporal dementia. *Neurology*, 56, S11–S15.

Myers, D. H., and Davies, P. (1978). The seasonal incidence of mania and its relationship to climatic variables. *Psychological Medicine*, 8, 433–440.

Nagano-Saito, A., Kato, T., Arahata, Y., Washimi, Y., Nakamura, A., Abe, Y., Yamada, T., Iwai, K., Hatano, K., Kawasumi, Y., Kachi, T., Dagher, A., and Ito, K. (2004). Cognitive- and motor-related regions in Parkinson's disease: FDOPA and FDG PET studies. *Neuroimage*, 22, 553–561.

Nagaraj, R., Singhi, P., and Malhi, P. (2006). Risperidone in children with autism: randomized, placebo-controlled, double-blind study. *Journal of Child Neurology*, 21, 450–455.

Nagoshi, C. T., and Johnson, R. C. (1987). Between- vs. within-family analyses of the correlation of height and intelligence. *Social Biology*, 34, 110–113.

Nakasato, A., Nakatani, Y., Seki, Y., Tsujino, N., Umino, M., and Arita, H. (2008). Swim stress exaggerates the hyperactive mesocortical dopamine system in a rodent model of autism. *Brain Research*, 1193, 128–135.

Narrow, W. E., Rae, D. S, Robins, L. N., and Regier, D. A. (2002). Revised prevalence estimates of mental disorders in the United States: using a clinical significance criterion to reconcile 2 surveys' estimates. *Archives of General Psychiatry*, 59, 115–123.

Nasar, S. (1998). *A Beautiful Mind: the Life of Mathematical Genius and Nobel Laureate John Nash*. New York: Touchstone.

Nathanielsz, P. W. (1999). *Life in the Womb: the Origin of Health and Disease*. Ithaca: Promethean.

National Science Foundation (1996). *Federal Scientists and Engineers: 1989–1993* (NSF 95–336). Washington, DC: National Science Foundation.

National Science Foundation (1999). *Scientist and Engineers Statistical Data System* (SESTAT). http://srsstats.sbe.nsf.gov.

Nayate, A., Bradshaw, J. L., and Rinehart, H. J. (2005). Autism and Asperger's disorder: are they movement disorders involving the cerebellum and/or basal ganglia? *Brain Research Bulletin*, 67, 327–334.

Nebes, R. D. (1974). Hemispheric specialization in commissurotomized man. *Psychological Bulletin*, 81, 1–14.

Nelson, E. E., and Panksepp, J. (1998). Brain substrates of infant–mother attachment: contributions of opioids, oxytocin, and norepinephrine. *Neuroscience and Biobehavioral Reviews*, 22, 437–452.

Nelson, J. C., Portera, L., and Leon, A. C. (2005). Are there differences in the symptoms that respond to a selective serotonin or norepinephrine reuptake inhibitor? *Biological Psychiatry*, 57, 1535–1542.

Newbury, D. F., Bonora, E., Lamb, J. A., Fisher, S. E., Lai, C. S., Baird, G., Jannoun, L., Slonims, V., Stott, C. M., Merricks, M. J., Bolton, P. F., Bailey, A. J., and Monaco, A. P., International Molecular Genetic Study of Autism Consortium. (2002). FOXP2 is not a major susceptibility gene for autism or specific language impairment. *American Journal of Human Genetics*, 70, 1318–1327.

Nieoullon, A. (2002). Dopamine and the regulation of cognition and attention. *Progress in Neurobiology*, 67, 53–83.

Niess, A. M., Feherenbach, E., Roecker, K., Lehmann, R., Opavsky, L., and Dickhuth, H. H. (2003). Individual differences in self-reported heat tolerance. Is there a link to the cardiocirculatory, thermoregulatory and hormonal response to endurance exercise in heat. *Sports Medicine and Physical Fitness*, 43, 386–392.

Nomura, Y., and Segawa, M. (2003). Neurology of Tourette's syndrome (TS) TS as a developmental dopamine disorder: a hypothesis. *Brain and Development*, 25 (Suppl. 1), S37–S42.

Noonan, J. P., Coop, G., Kudaravalli, S., Smith, D., Krause, J., Alessi, J., Chen, F., Platt, D., Paabo, S., Pritchard, J. K., and Rubin, E. M. (2006).

Sequencing and analysis of Neanderthal genomic DNA. *Science*, 314, 1068–1071.

Noshirvani, H. F., Kasvikis, Y., Marks, I. M., Tsakiris, F., and Monteiro, W. O., (1991). Gender-divergent aetiological factors in obsessive-compulsive disorder. *British Journal of Psychiatry*, 158, 260–263.

Nunneley, S. A. (1996). Thermal stress. In R. L. Dehart (ed.), *Fundamentals of Aerospace Medicine* (2nd edn) (pp. 399–422). Baltimore: Williams & Wilkins.

O'Brien, J. M. (1992). *Alexander the Great: the Invisible Enemy*. London: Routledge.

O'Connell, J. F. (2007). How did modern humans displace Neanderthals? Insights from hunter-gatherer ethnography and archaeology. In N. Conard (ed.), *Neanderthals and Modern Humans Meet* (pp. 43–64). Tübingen: Kerns Verlag.

O'Donovan, D. K. (1996). Hypothyroid nails and evolution. *Lancet*, 347, 1261–1262.

Okuda, J., Fujii, T., Ohtake, H., Tsukiura, T., Tanji, K., Suzuki, K., Kawashima, R., Fukuda, H., Itoh, M., and Yamadori, A. (2003). Thinking about the future and past: the roles of the frontal pole and the medial temporal lobes. *Neuroimage*, 19, 1369–1380.

Oldenziel, R. (2004). *Making Technology Masculine: Men, Women and Modern Machines in America, 1870–1945*. Amsterdam: Amsterdam University Press.

Ollat, H. (1992). Dopaminergic insufficiency reflecting cerebral ageing: value of a dopaminergic agonist, piribedil. *Journal of Neurology*, 239, S13–S16.

Olmstead, D. (2005). The age of autism: Amish ways. *United Press International*, June 7, 2005 (www.washtimes.com/upi-breaking/20050606–100328–8006r.htm).

Ongur, D., and Price, J. L. (2000). The organization of networks within the orbital and medial prefrontal cortex of rats, monkeys and humans. *Cerebral Cortex*, 10, 206–219.

Ornstein, R. E. (1972). *The Psychology of Consciousness*. San Francisco: W. H. Freeman.

Oskamp, S. (2000). A sustainable future for humanity? How can psychology help? *American Psychologist*, 55, 496–508.

Otto, M. W. (1992). Normal and abnormal information processing. A neuropsychological perspective on obsessive compulsive disorder. *Psychiatric Clinics of North America*, 15, 825–848.

Ozonoff, S., and Miller, J. N. (1996). An exploration of right-hemisphere contributions to the pragmatic impairments of autism. *Brain and Language*, 52, 411–434.

Padavic, I., and Reskin, B. F. (2003). *Women and Men at Work* (2nd edn). Thousand Oaks: Pine Forge Press.

Pahnke, W. N. (1969). Psychedelic drugs and mystical experience. *International Psychiatry Clinics*, 5, 149–162.

Palit, G., Kumar, R., Gupta, M. B., Saxena, R. C., Patnaik, G. K., and Dhawan, B. N. (1997). Quantification of behaviour in social colonies of rhesus monkey. *Indian Journal of Physiology and Pharmacology*, 41, 219–226.

Palmer, R. F., Blanchard, S., Stein, Z., Mandell, D., and Miller, C. (2006). Environmental mercury release, special education rates, and autism disorder: an ecological study of Texas. *Health and Place*, 12, 203–209.

Pani, L. (2000). Is there an evolutionary mismatch between the normal physiology of the human dopaminergic system and current environmental conditions in industrialized countries? *Molecular Psychiatry*, 5, 467–475.

Panksepp, J. (1999). The affiliative playfulness and impulsivity of extraverts may not be dopaminergically mediated. *Behavioral and Brain Sciences*, 22, 533–534.

Papolas, D. F., and Papolas, J. (2002). *The Bipolar Child* (rev. edn). New York: Broadway Books.

Paradiso, S., Robinson, R. G., and Arndt, S. (1996). Self-reported aggressive behavior in patients with stroke. *Journal of Nervous and Mental Disease*, 184, 746–753.

Parke, B. (2003). *Einstein: the Passions of a Scientist*. New York: Prometheus Books.

Parsons, S., Mitchell, P., and Leonard, A. (2004). The use and understanding of virtual environments by adolescents with autistic spectrum disorders. *Journal of Autism and Developmental Disorders*, 34, 449–466.

Pelham, W. E., Murphy, D. A., Vannatta, K., Milich, R., Licht, B. G., Gnagy, E. M., Greenslade, K. E., Greiner, A. R., and Vodde-Hamilton, M. (1992). Methylphenidate and attributions in boys with attention-deficit hyperactivity disorder. *Journal of Consulting and Clinical Psychology*, 60, 282–292.

Pepperberg. I. (1990). Conceptual abilities of some nonprimate species, with an emphasis on an African Grey parrot. In S. T. Parker and K. R. Gibson (eds), *"Language" and Intelligence in Monkeys and Apes* (pp. 459–407). New York: Cambridge University Press.

Perry, E. K., and Perry, R. H. (1995). Acetylcholine and hallucinations: disease-related compared to drug-related alterations in human consciousness. *Brain and Cognition*, 28, 240–258.

Perry, R. J., Rosen, H. R., Kramer, J. H., Beer, J. S., Levenson, R. L., and Miller, B. L. (2001). Hemispheric dominance for emotions, empathy and social behaviour: evidence from right and left handers with frontotemporal dementia. *Neurocase*, 7, 145–160.

Persinger, M. A. (1984). People who report religious experiences may also display enhanced temporal-lobe signs. *Perceptual and Motor Skills*, 58, 963–975.

Persinger, M. A., and Fisher, S. D. (1990). Elevated, specific temporal lobe signs in a population engaged in psychic studies. *Perceptual and Motor Skills*, 71, 817–818.

Persinger, M. A., and Makarec, K. (1987). Temporal lobe epileptic signs and correlative behaviors displayed by normal populations. *Journal of General Psychology*, 114, 179–195.

Peters, M., Reimers, S., and Manning, J. T. (2006). Hand preference for writing and associations with selected demographic and behavioral variables in 255,100 subjects: the BBC internet study. *Brain and Cognition*, 62, 177–189.

Petronis, A. (2001). Human morbid genetics revisited: relevance of epigenetics. *Trends in Genetics*, 17, 142–146.

Petty, F., Davis, L. L., Kabel, D., and Kramer, G. L. (1996). Serotonin dysfunction disorders: a behavioral neurochemistry perspective. *Journal of Clinical Psychiatry*, 57 (Suppl. 8), 11–16.

Pickett, E. R., Kuniholm, E., Protopapas, A., Friedman, J., and Lieberman, P. (1998). Selective speech motor, syntax and cognitive deficits associated with bilateral damage to the putamen and the head of the caudate nucleus: a case study. *Neuropsychologia*, 6, 173–188.

Pierce, R. C., Crawford, C. A., Nonneman, A. J., Mattingly, B. A., and Bardo, M. T. (1990). Effect of forebrain dopamine depletion on novelty-induced place preference behavior in rats. *Pharmacology, Biochemistry and Behavior*, 36, 321–325.

Pierrot-Deseilligny, C., Muri, R. M., Ploner, C. J., Gaymard, B., and Rivaud-Pechoux, S. (2003). Cortical control of ocular saccades in humans: a model for motricity. *Progress in Brain Research*, 142, 3–17.

Pillmann, F., Rohde, A., Ullrich, S., Draba, S., Sannemuller, U., and Marneros, A. (1999). Violence, criminal behavior, and the EEG: significance of left hemispheric focal abnormalities. *Journal of Neuropsychiatry and Clinical Neuroscience*, 11, 454–457.

Pinker, S., and Bloom, P. (1990). Natural language and natural selection. *Behavioral and Brain Sciences*, 13, 707–784.

Pliszka, S. R. (2005). The neuropsychopharmacology of attention-deficit/hyperactivity disorder. *Biological Psychiatry*, 57, 1385–1390.

Poizner, H., and Kegl, J. (1993). Neural disorders of the linguistic use of space and movement. *Annals of the New York Academy of Sciences*, 682, 192–213.

Poizner, H., Fookson, O. I., Berkinblit, M. B., Hening, W., Feldman, G., and Adamovich, S. (1998). Pointing to remembered targets in 3-D space in Parkinson's disease. *Motor Control*, 2, 251–277.

Prabhakaran, V., Smith, J. A. L., Desmond, J. E., Glover, G. H., and Gabrieli, J. D. E. (1997). Neural substrates of fluid reasoning: an fMRI study of neocortical activation during performance of the Raven's Progressive Matrices Test. *Cognitive Psychology*, 33, 43–63.

Prather, J. F., and Mooney, R. (2004). Neural correlates of learned song in the avian forebrain: simultaneous representation of self and others. *Current Opinion in Neurobiology*, 14, 496–502.

Prescott, C. A., Johnson, R. C., and McArdle, J. J. (1999). Chorion type as a possible influence on the results and interpretation of twin study data. *Twin Research*, 2, 244–249.

Previc, F. H. (1991). A general theory concerning the prenatal origins of cerebral lateralization in humans. *Psychological Review*, 98, 299–334.

(1993) Do the organs of the labyrinth differentially influence the sympathetic and parasympathetic systems? *Neuroscence and Biobehavioral Reviews*, 17, 397–404.

(1996). Nonright-handedness, central nervous system and related pathology, and its lateralization: a reformulation and synthesis. *Developmental Neuropsychology*, 12, 443–515.

(1998). The neuropsychology of 3-D space. *Psychological Bulletin*, 124, 123–164.

(1999). Dopamine and the origins of human intelligence. *Brain and Cognition*, 41, 299–350.

(2002). Thyroid hormone production in chimpanzees and humans: implications for the origins of human intelligence. *American Journal of Physical Anthropology*, 118, 402–403.

(2004). An integrated neurochemical perspective on human performance measurement. In J. W. Ness, V. Tepe and D. R. Ritzer (eds), *The Science and Simulation of Human Performance* (pp. 327–390). Amsterdam: Elsevier.

(2006). The role of the extrapersonal brain systems in religious activity. *Consciousness and Cognition*, 15, 500–539.

(2007). Prenatal influences on brain dopamine and their relevance to the rising incidence of autism. *Medical Hypotheses*, 68, 46–60.

Previc, F. H., Declerck, C., and De Brabander, B. (2005). Why your "head is in the clouds" during thinking: the relationship between cognition and upper space. *Acta Psychologia*, 118, 7–24.

Primavesi, A. (1991). *From Apocalypse to Genesis: Ecology, Feminism, and Christianity*. Minneapolis: Fortress Press.

Purdon, S. E., Chase, T., and Moehr, E. (1996). Huntington's disease or chorea. In J. G. Beaumont, P. M. Kenealy and M. J. C. Rogers (eds), *The Blackwell Dictionary of Neuropsychology* (pp. 401–406). Cambridge, MA: Blackwell Publishers.

Pycock, C. J., Donaldson, I. M., and Marsden, C. D. (1975). Circling behaviour produced by unilateral lesions of the locus coeruleus in rats. *Brain Research*, 97, 317–329.

Quadri, R., Comino, I., Scarzella, L., Cacioli, P., Zanone, M. M. , Pipieri, A., Bergamasco, B., and Chiandussi, L. (2000). Autonomic nervous function in de novo parkinsonian patients in basal condition and after acute levodopa administration. *Functional Neurology*, 15, 81–86.

Rakic, P. (1996). Evolution of neocortical parcellation: the perspective from experimental neuroembryology. In J.-P. Changeux and J. Chavaillon (eds), *Origins of the Human Brain* (pp. 84–100). Oxford: Clarendon.

Rapoport, S. I. (1990). Integrated phylogeny of the primate brain, with special reference to humans and their diseases. *Brain Research Reviews*, 15, 267–294.

Reeves, S. J., Grasby, P. M., Howard, R. J., Bantick, R. A., Asselin, M. C., and Mehta, M. A. (2005). A positron emission tomography (PET) investigation of the role of striatal dopamine (D2) receptor availability in spatial cognition. *Neuroimage*, 28, 216–226.

Regehr, C., Hill. J., and Glancy, G. D. (2000). Individual predictors of traumatic reactions in firefighters. *Journal of Nervous and Mental Disease*, 188, 333–339.

Reiss, M., Tymnik, G., Kogler, P., Kogler, and Reiss, G. (1999). Laterality of hand, foot, eye, and ear in twins. *Laterality*, 4, 287–297.

Ressler, K. J., Sullivan, S. L., and Buck, L. B. (1994). A molecular dissection of spatial patterning in the olfactory system. *Current Opinion in Neurobiology*, 4, 588–596.

Reuter, M., Panksepp, J., Schnabel, N., Kellerhoff, N., Kempel, P., and Hennig, J. (2005). Personality and biological markers of creativity. *European Journal of Personality*, 19, 83–95.

Richards, M. P., Petitt, P. B., Stiner, M. C., and Trinkaus, E. (2001). Stable isotope evidence for increasing dietary breadth in the European mid-Upper Paleolithic. *Proceedings of the National Academy of Sciences*, 98, 6528–6532.

Richman, D. P., Stewart, R. M., Hutchinson, J. W., and Caviness, V. S. Jr. (1975). Mechanical model of brain convolutional development. *Science*, 189, 18–21.

Ridley, R. M., and Baker, H. F. (1982). Stereotypy in monkeys and humans. *Psychological Medicine*, 12, 61–72.

Rihet, P., Possamai, C. A., Micallef-Roll, J., Blin, O., and Hasbroucq, T. (2002). Dopamine and human information processing: a reaction-time analysis of the effect of levodopa in healthy subjects. *Psychopharmacology*, 163, 62–77.

Robbins, T. W. (2000). Chemical neuromodulation of frontal-executive functions in humans and other animals. *Experimental Brain Research*, 133, 130–138.

Robbins, T. W., and Everitt, B. J. (1982). Functional studies of the central catecholamines. *International Review of Neurobiology*, 23, 303–365.

Robertson, M. M. (2003). Diagnosing Tourette syndrome: is it a common disorder? *Journal of Psychosomatic Research*, 55, 3–6.

Robinson, P. D., Schutz, C. K., Macciardi, F., White, B. N., and Holden, J. J. A. (2001). Genetically determined low maternal serum dopamine β-hydro-xylase levels and the etiology of autism spectrum disorders. *American Journal of Medical Genetics*, 100, 30–36.

Robison, L. M., Skaer, T. L., Sclar, D. A., and Galin, R. S. (2002). Is attention deficit hyperactivity disorder increasing among girls in the US? Trends in diagnosis and the prescribing of stimulants. *CNS Drugs*, 16, 129–137.

Rodriguez, P. F., Aron, A. R., and Poldrack, R. A. (2006). Ventral-striatal/nucleus-accumbens sensitivity to prediction errors during classification learning. *Human Brain Mapping*, 27, 306–313.

Rosenberg, D. R., and Keshavan, M. S. (1998). Toward a neurodevelopmental model of obsessive-compulsive disorder. *Biological Psychiatry*, 43, 623–640.

Rosenzweig, M. R., Breedlove, S. M., and Leiman, A. L. (2002). *Biological psychology* (3rd edn). Sunderland, MA: Sinauer Associates.

Rosse, R. B., Collins, J. P., Jr., Fay-Mccarthy, M., Alim, T. N., Wyatt, R. J., and Deutsch, S. I. (1994). Phenomenologic comparison of the idiopathic psychosis of schizophrenia and drug-induced cocaine and phencyclidine psychoses: a retrospective study. *Clinical Neuropharmacology*, 17, 359–369.

Rosvold, H. E., and Mishkin, M. (1950). Evaluation of the effects of prefrontal lobotomy on intelligence. *Canadian Journal of Psychology*, 4, 122–126.

Rotenberg, V. S. (1994). An integrative psychophysiological approach to brain hemisphere functions in schizophrenia. *Neuroscience and Biobehavioral Reviews*, 18, 487–395.

Ruiz-Lozano, P., Ryan, A. K., and Izpisua-Belmonte, J. C. (2000). Left-right determination. *Trends in Cardiovascular Medicine*, 10, 258–262.

Russell, A. J., Mataix-Cols, D., Anson, M., and Murphy, D. G. (2005). Obsessions and compulsions in Asperger syndrome and high-functioning autism. *British Journal of Psychiatry*, 186, 525–528.

Sabbagh, M. A. (1999). Communicative intentions and language: evidence from right-hemisphere damage and autism. *Brain and Language*, 70, 29–69.

Sachdev, P. (1998). Schizophrenia-like psychosis and epilepsy: the status of the association. *American Journal of Psychiatry*, 155, 325–336.

Sahlin, M. (1972). *Stone Age Economics*. Chicago: Aldine and Atherton, Inc.

Sakai, L. M., Baker, L. A., Jacklin, C. N., and Shulman, I. (1991). Sex steroids at birth: genetic and environmental variation and covariation. *Developmental Psychobiology*, 24, 559–570.

Salamone, J. D., and Correa, M. (2002). Motivational views of reinforcement: implications for understanding the behavioral functions of nucleus accumbens dopamine. *Behavioral and Brain Research*, 137, 3–25.

Salamone, J. D., Correa, M., Mingote, S. M., and Weber, S. M. (2005). Beyond the reward hypothesis: alternative functions of nucleus accumbens dopamine. *Current Opinion in Pharmacology*, 5, 34–41.

Santana, C., Martin, L., and Rodriguez-Diaz, M. (1994). Tyrosine ingestion during rat pregnancy alters postnatal development of dopaminergic neurons in the offspring. *Brain Research*, 635, 96–102.

Sanua, V. D. (1983). Infantile autism and childhood schizophrenia: review of the issues from the sociocultural point of view. *Social Science in Medicine*, 17, 1633–1651.

Sartorius, N., Jablensky, A. Korten, A., Ernberg, G., Anker, M., Cooper, J. E., and Day, R. (1986). Early manifestations and first-contact incidence of schizophrenia in different cultures. A preliminary report on the initial evaluation phase of the WHO Collaborative Study on determinants of outcome of severe mental disorders. *Psychological Medicine*, 16, 909–928.

Saver, J. L., and Rabin, J. (1997). The neural substrates of religious experience. *Journal of Neuropsychiatry and Clinical Neurosciences*, 9, 498–510.

Sawamoto, N., Honda, M., Hanakawa , T., Fukuyama, H., and Shibasaki, H. (2002). Cognitive slowing in Parkinson's disease: a behavioral evaluation independent of motor slowing. *Journal of Neuroscience*, 22, 5198–5203.

Scherer, D. M. (2001). Adverse perinatal outcome of twin pregnancies according to chorionicity: review of the literature. *American Journal of Perinatology*, 18, 23–37.

Schlemmer, R. F., Jr., Narasimhachari, N., and Davis, J. M. (1980). Dose-dependent behavioural changes induced by apomorphine in selected members of a primate social colony. *Journal of Pharmacy and Pharmacology*, 32, 285–289.

Schmitt, J. A., Rameakers, J. G., Kruizinga, M. J., Van Bostel, M. P., Vuurman, E. F., and Reidel, W. J. (2002). Additional dopamine reuptake inhibition attenuates vigilance impairment induced by serotonin reuptake inhibition in man. *Journal of Psychopharmacology*, 16, 207–214.

Schoenemann, P. T., Budinger, T. F., Sarich, V. M., and Wang, W. S. (2000). Brain size does not predict general cognitive ability within families. *Proceedings of the National Academy of Sciences*, 97, 4932–4937.

Schuck, S., Bentue-Ferrer, D., Kleinermans, D., Reymann, J.-M., Polard, E., Gandon, J. M., and Allain, H. (2002). Psychomotor and cognitive effects of piribedil, a dopamine agonist, in young healthy volunteers. *Fundamental and Clinical Pharmacology*, 16, 57–65.

Schultz, W., Dayan, P., and Montague, R. (1997). A neural substrate of prediction and reward. *Science*, 275, 1593 –1599.

Schwart, R. G., Uretsky, N. J., and Bianchine, J. R. (1982). Prostaglandin inhibition of amphetamine-induced circling in mice. *Psychopharmacology*, 78, 317–321.

Schwartz, B. (2004). The tyranny of choice. *Scientific American*, 290(4), 70–75.

Seligman, M. (1975). *Helplessness*. San Fransisco: Freeman.

Semendeferi, K., Lu, A., Schenker, N., and Damasio, H. (2002). Humans and great apes share a large frontal cortex. *Nature Neuroscience*, 5, 272–276.

Senghas, A., Kita, S., and Ozyurek, A. (2004). Children creating core properties of language: evidence from an emerging sign language in Nicaragua. *Science*, 305, 1779–1782.

Sevy, S., Hassoun, Y., Bechara, A., Yechiam, E., Napolitano, B., Burdick, K., Delman, H., and Malhotra, A. (2006). Emotion-based decision-making in healthy subjects: short-term effects of reducing dopamine levels. *Psychopharmacology*, 188, 228–235.

Shapiro, T., and Hertzig, M. E. (1991). Social deviance in autism: a central integrative failure as a model for social nonengagement. *Psychiatric Clinics of North America*, 14, 19–32.

Shea, J. J. (2003). Neanderthals, competition, and the origin of modern human behavior in the Levant. *Evolutionary Anthropology*, 12, 173–187.

Shipman, P. (1986). Scavenging or hunting in early hominids. Theoretical framework and tests. *American Anthropologist*, 88, 27–43.

Sicotte, N. L, Woods, R. P., and Mazziotta, J. C. (1999). Handedness in twins: a meta-analysis. *Laterality*, 4, 265–286.

Sierra, M., and Berrios, G. E. (1998). Depersonalization: neurobiological perspectives. *Biological Psychiatry*, 44, 898–908.

Siever, L. J. (1994) Biologic factors in schizotypal personal disorders. *Acta Psychiatrica*, 384, 45–50.

Simon, H., Scatton, B., and Le Moal, M. (1980). Dopaminergic A10 neurones are involved in cognitive functions. *Nature*, 286, 150–151.

Simonton, D. K. (1994). *Greatness: Who Makes History and Why*. New York: Guilford Press.

Skoyles, J. R. (1999). Human evolution expanded brains to increase expertise capacity, not IQ. *Psycholoquy*, 10(002).

Smith, A., and Sugar, O. (1975). Development of above normal language and intelligence 21 years after left hemispherectomy *Neurology*, 25, 813–818.

Smith, A. B., Taylor, E., Brammer, M., and Rubia, K. (2004). Neural correlates of switching set as measured in fast, event-related functional magnetic resonance imaging. *Human Brain Mapping*, 21, 247–256.

Solms, M., (2000). Dreaming and REM sleep are controlled by different brain mechanisms. *Behavioral and Brain Sciences*, 23, 793–1121.

Spivak, M., and Epstein, M. (2001). Newton's psychosis. *American Journal of Psychiatry*, 158, 821–822.

Stahlberg, O., Soderstrom, H., Rastam, M., and Gillberg, C. (2004). Bipolar disorder, schizophrenia, and other psychotic disorders in adults with childhood onset AD/HD and/or autism spectrum disorders. *Journal of Neural Transmission*, 111, 891–902.

Stedman, H. H., Kozyak, B. W., Nelson, A., Thesier, D. M., Su, L. T., *et al.*
(2004). Myosin gene mutation correlates with anatomical changes in the
human lineage. *Nature*, 428, 415–418.

Stenger, R. (2002). Study: universe could end in 10 billion years, *CNN*,
September 18, 2002 (http://archives.cnn.com/2002/TECH/space/09/18/
cosmic.crunch/index.html).

Stern, P. C. (2000). Psychology and the science of human-environment interactions. *American Psychologist*, 55, 523–530.

Stromland, K., Nordin, V., Miller, M., Akerstrom, B., and Gillberg, C. (1994).
Autism in thalidomide embryopathy: a population study. *Developmental
Medicine and Child Neurology*, 36, 351–356.

Suddendorf, T., and Corballis, M. (1997). Mental time travel and the evolution
of the human mind. *Genetic, Social and General Psychology Monographs*, 123,
133–167.

Sundberg, N. D., Poole, M. E., and Tyler, L. E. (1983). Adolescents' expect-
ations of future events: a cross-cultural study of Australians, Americans, and
Indians. *International Journal of Psychology*, 18, 415–427.

Sutoo, D., and Akiyama, K. (1996). The method by which exercise modifies
brain function. *Physiology and Behavior*, 60, 177–181.

Suzuki, T. (1981). How great will the stature of Japanese eventually become?
Journal of Human Ergology, 10, 13–24.

Svebak, S. (1985). Serious-mindedness and the effect of self-induced respiratory
changes upon parietal EEG. *Biofeedback and Self-Regulation*, 10, 49–62.

Swerdlow, N. R., and Koob, G. F. (1987). Dopamine, schizophrenia, mania and
depression: toward a unified hypothesis of cortico-striatal-pallidothalamic
function. *Behavioral and Brain Sciences*, 10, 197–245.

Szechtman, H., Talangbayan, H., and Eilam, D. (1993). Environmental and
behavioral components of sensitization induced by the dopamine agonist
quinpirole. *Behavioral Pharmacology*, 4, 405–410.

Szechtman, H., Culver, K., and Eilam, D. (1999). Role of dopamine systems
in obsessive-compulsive disorder (OCD): implications from a novel
psychostimulant-induced animal model. *Polish Journal of Pharmacology*, 51,
55–61.

Tallal, P., Miller, S., and Fitch, R. H. (1993). Neurobiological basis of speech: a
case for the preeminence of temporal processing. *Annals of the New York
Academy of Sciences*, 682, 27–47.

Tanaka, S., Kanzaki, R., Yoshibayashi, M., Kamiya, T., and Sugishita, M.
(1999). Dichotic listening in patients with situs inversus: brain asymmetry
and situs asymmetry. *Neuropsychologia*, 37, 869–874.

Taylor, C. R., and Rowntree, V. J. (1973). Temperature regulation and heat
balance in running cheetahs: a strategy for sprinters? *American Journal of
Physiology*, 224, 848–851.

Taylor, S. (2005). *The Fall: the Evidence for a Golden Age, 6,000 Years of Insanity
and the Dawning of a New Era* Oakland, CA: O Books.

Tekin, S., and Cummings, J. L. (2002). Frontal-subcortical neuronal circuits
and clinical neuropsychiatry: an update. *Journal of Psychosomatic Research*,
53, 647–654.

Templeton, A. R. (2002). Out of Africa again and again. *Nature*, 416, 45–51.

Thomas, S. A., Matsumoto, A. M., and Palmiter, R. D. (1995). Noradrenaline is essential for mouse fetal development. *Nature*, 374, 643–646.

Todorov, I. T. (2005). Einstein and Hilbert: the creation of general relativity. Colloquium lecture at the International University Bremen, March 15, 2005 (http://arxiv.org/PS_cache/physics/pdf/0504/0504179.pdf).

Toffler, A. (1970). *Future Shock*. New York: Random House.

Torrey, E. F., Miller, J., Rawlings, R., and Yolken, R. H. (1997). Seasonality of births in schizophrenia and bipolar disorder: a review of the literature. *Schizophrenia Research*, 28, 1–38.

Toth, N. (1985). Archaeological evidence for preferential right-handedness in the lower and middle Pleistocene and its possible implications. *Journal of Human Evolution*, 14, 607–614.

Truett, G. E., Brock, J. W., Lidl, G. M., and Kloster, C. A. (1994). Stargazer (stg), new deafness mutant in the Zucker rat. *Laboratory Animal Science*, 44, 595–599.

Tucker, D. M., and Williamson, P. A. (1984). Asymmetric neural control systems in human self-regulation. *Psychological Review*, 91, 185–215.

Tucker, T. J. and Kling, A. (1969). Preservation of delayed response following combined lesions of prefrontal and posterior association cortex in infant monkeys. *Experimental Neurology*, 23, 491–502.

U. S. Census Bureau. (2004). *2000 Census of Population and Housing: Population and Housing Counts* (Pt 1). Washington, DC, Department of Commerce.

Van Schaik, C. (2006). Why are some animals so smart? *Scientific American*, 294 (4), 64–71.

Vanhaereny, M., d'Errico, F., Stringer, C., James, S. L., Todd, J. A., and Mienis, H. K. (2006). Middle Paleolithic shell beads in Israel and Algeria. *Science*, 312, 1785–1788.

Vargha-Khadem, F., Watkins, K., Alcock, K., Fletcher, P., and Passingham, R. (1995). Praxic and nonverbal cognitive deficits in a large family with a genetically transmitted speech and language disorder. *Proceedings of the National Academy of Sciences*, 92, 930–933.

Vedantum, S. (2004). Antidepressant use of U.S. adults soars. *Washington Post*, December 3, 2004 (www.washingtonpost.com/wp-dyn/articles/A29751–2004Dec2.html).

Vernaleken, I., Weibrich, C., Siessmeier, T., Buchholz, H. G., Rosch, F., Heinz, A., Cumming, P., Stoeter, P., Bartenstein, P., and Grunder, G. (2007). Asymmetry in dopamine D(2/3) receptors of caudate nucleus is lost with age. *Neuroimage*, 34, 870–878.

Viggiano, D., Vallone, D., Ruocco, L. A., and Sadile, A. G. (2003). Behavioural, pharmacological, morpho-functional molecular studies reveal a hyperfunctioning mesocortical dopamine system in an animal model of attention deficit and hyperactivity disorder. *Neuroscience and Biobehavioral Reviews*, 27, 683–689.

Villardita, C. (1985). Raven's colored progressive matrices and intellectual impairment in patients with focal brain damage. *Cortex*, 21, 627–634.

Vingerhoets, G., de Lange, F. P., Vandemaele, P., Deblaere, K., and Achten, E. (2002). Motor imagery in mental rotation: an fMRI study. *Neuroimage*, 17, 1623–1633.

Volkmar, F. R. (2001). Pharmacological interventions in autism: theoretical and practical issues. *Journal of Clinical Child Psychology*, 30, 80–87.

Volkow, N. D., Gur, R. C., Wang, G.-J., Fowler, J. S., Moberg, P. J., Ding, Y.-S., Hitzemann, R., Smith, G., and Logan, J. (1998). Association between decline in brain dopamine activity with age and cognitive and motor impairment in healthy individuals. *American Journal of Psychiatry*, 155, 344–349.

Volkow, N. D., Wang, G. J., Ma, Y., Fowler, J. S., Wong, C., Ding, Y. S., Hitzemann, R., Swanson, J. M., and Kalivas, P. (2005). Activation of orbital and medial prefrontal cortex by methylphenidate in cocaine-addicted subjects but not in controls: relevance to addiction. *Journal of Neuroscience*, 25, 3932–3939.

Vollenweider, F. X., and Geyer, M. A. (2001). A systems model of altered consciousness: integrating natural and drug-induced psychoses. *Brain Research Bulletin*, 56, 495–507.

Von Haesler, A., Sajantila, A., and Paabo, S. (1996). The genetical archaeology of the human genome. *Nature Genetics*, 14, 135–140.

Vuilleumier, P., Ortigue, S., and Brugger, P. (2004). The number space and neglect. *Cortex*, 40, 399–410.

Wainwright, P. E. (2002). Dietary essential fatty acids and brain function: a developmental perspective on mechanisms. *Proceedings of the Nutrition Society*, 61, 61–69.

Waldmann, C., and Gunturkun, C. (1993). The dopaminergic innervation of the pigeon caudolateral forebrain: immunocytochemical evidence for a "prefrontal cortex" in birds? *Brain Research*, 600, 225–234.

Walter, R. C., Buffer, R. T., Bruggemann, J. H., Guillaume, M. M. M., Berhe, S. M., Negassi, B., Libsekal, Y., Cheng, H., Edwards, R. W., Von Cosel, R., Neraudeau, D., and Gagnon, M. (2000). Early human occupation of the Red Sea coast of Eritrea during the last interglacial. *Nature*, 405, 65–69.

Wanpo, H., Clochon, R., Yumin, G., Larick, R., Qiren, F., Schwarcz, H., Yonge, C., de Vos, J., and Rink, W. (1995). Early Homo and associated artefacts from Asia. *Nature*, 378, 275–278.

Watson, J. B., Mednick, S. A., Huttunen, M., and Wang, X. (1999). Prenatal teratogens and the development of adult mental illness. *Developmental Psychopathology*, 11, 457–466.

Watts, J. (2002). Public health experts concerned about "hikikomori". *Lancet*, 359, 1131.

Waziri, R. (1980). Lateralization of neuroleptic-induced dyskinesia indicates pharmacologic asymmetry in the brain. *Psychopharmacology*, 68, 51–53.

Webb, D. M., and Zhang, J. (2005). FoxP2 in song-learning birds and vocal-learning mammals. *Journal of Heredity*, 96, 212–216.

Weiner, I. (2003). The "two-headed" latent inhibition model of schizophrenia: modeling positive and negative symptoms and their treatment. *Psychopharmacology*, 169, 257–297.

Weintraub, S. and Mesulam, M. M. (1983). Developmental learning disabilities of the right hemisphere. Emotional, interpersonal, and cognitive components. *Archives of Neurology*, 40, 463–468.

Weiss, H., and Bradley, R. S. (2001). What drives social collapse? *Science*, 291, 609–610.

Welsh, M. C., Pennington, B. F., Ozonoff, S., Rouse, B., and McCabe, E. R. B. (1990). Neuropsychology of early-treated phenylketonuria: specific executive function deficits. *Child Development*, 61, 1697–1713.

Wheeler, P. E. (1985). The loss of functional body hair in humans: the influence of thermal environment, body form and bipedality. *Journal of Human Evolution*, 14, 23–28.

Whishaw, I. Q. (1993). Activation, travel distance, and environmental change influence food carrying in rats with hippocampal, medial thalamic and septal lesions: implications for studies on hoarding and theories of hippocampal function. *Hippocampus*, 3, 373–385.

Whishaw, I. Q., and Dunnett, S. B. (1985). Dopamine depletion, stimulation or blockade in the rat disrupts spatial navigation and locomotion dependent upon beacon or distal cues. *Behavioural Brain Research*, 18, 11–29.

Whiten, A. (1990). Cause and consequences in the evolution of hominid brain size. *Behavioral and Brain Sciences*, 13, 367.

Wilford, J. N. (1991). *The Mysterious History of Columbus*. New York: Knopf.

Wilkins, W. K., and Wakefield, J. 1995). Brain evolution and neurolinguistic preconditions. *Behavioral and Brain Sciences*, 18, 161–182.

Wilson, C. (1985). *A Criminal History of Mankind*. London: Grafton.

Winstanley, C. A., Theobald, D. E. H., Dalley, J. W., and Robbins, T. W. (2005). Interactions between serotonin and dopamine in the control of impulsive choice in rats: therapeutic implications of impulse control disorders. *Neuropsychopharmacology*, 30, 669–682.

Wittling, W., Block, A., Schweiger, E., and Genzel, S. (1998). Hemispheric asymmetry in sympathetic control of the human myocardium. *Brain and Cognition*, 38, 17–35.

Wolford, G., Miller, M. B., and Gazzaniga, M. (2000). The left hemisphere's role in hypothesis formation. *Journal of Neuroscience*, 20 (RC64), 1–4.

Woods, R. P. (1986). Brain asymmetries in situs inversus: a case report and review of the literature. *Archives of Neurology*, 43, 1083–1084.

Wynn, T., and Coolidge, F. L. (2004). The expert Neanderthal mind. *Journal of Human Evolution*, 46, 467–487.

Yates, B. J. (1996). Vestibular influences on the autonomic nervous system. *Annals of the New York Academy of Sciences*, 781, 458–473.

Yates, B. J., and Bronstein, A. M. (2005). The effects of vestibular system lesions on autonomic regulation: observations, mechanisms, and clinical implications. *Journal of Vestibular Research*, 15, 119–129.

Young, S. N., and Leyton, M. (2002). The role of serotonin in human mood and social interaction. Insight from altered tryptophan levels. *Pharmacology, Biochemistry, and Behavior*, 71, 857–865.

Zhang, J., Wang, L., and Pitts, D. K. (1996). Prenatal haloperidol reduces the number of active midbrain dopamine neurons in rat offspring. *Neurotoxicology and Teratology*, 18, 49–57.

Zhou, T. (1998). Energy consumption in the United States. In G. Elert (ed.), *The Physics Factbook* (www.hypertextbook.com/facts).

Zihlman, A. L., and Cohn, B. A. (1988). The adaptive response of human skin to the savanna. *Human Evolution*, 3, 397–409.

Zilhão, J., d'Errico, F., Bordes, J. G., Lenoble, A., Texier, J. P., and Rigaud, J. P. (2006). Analysis of Aurignacian interstratification at the Chatelperronian-type site and implications for the behavioral modernity of Neandertals. *Proceedings of the National Academy of Sciences*, 103, 12643–12648.

Zimbardo, P. (2002). Time to take our time; looking to the future is important – and very American – but living in the present is vital – just think about it. *Psychology Today*, March/April 2002.

Zuckerman, M. (1984). Sensation-seeking: a comparative approach to a human trait. *Behavioral and Brain Sciences*, 7, 413–471.

Index

!Kung San, 125
abstraction, 2–3
  and intelligence, 61–2
  dopamine involvement, 16
abulia, 29
acetylcholine, 14, 20, 21, 25, 56
achievement motivation, 69, 70, 79, 83, 166
action-extrapersonal brain system
  and ventromedial dopaminergic
      personality, 39–40
  extrapersonal experiences, 53
adynamia, 29
affilliative extraversion, 65
agentic extraversion, 65
agentic extrovert personality, 70
agrarian societies, 125–6
Alexander the Great, 134–5, 148–9
Alzheimer's disease, 16
ambient extrapersonal brain system,
    38–9
Amish society, absence of autism, 151
amphetamine, 20, 21, 23, 27, 82, 94
  maternal ingestion of, 102–3
amygdala, 24, 27
anosognosia, 53
anterior-posterior axis, 23
anti-fever agents, 111
Antikythera Mechanism, 127
antipsychotics, 23
anti-saccade, 60
apathy, 29
aphasias, 7, 8
apomorphine, 21
arts, diminishing returns in, 157
Asperger's syndrome, 81
attention-deficit disorder, 28, 68, 75
attention-deficit/hyperactivity disorder,
    79–81, *see also* hyperdopaminergic
    disorders
  excessive dopamine activity, 75
  generativity/creativity, 64
  obsessive-compulsive disorder, 87

prevalence, 76
  schizophrenia, 80, 91
Australopithecines, 104–5
Australopithecus, 104
Australopithecus afarenis, 104–5
autism, 81–3, *see also* hyperdopaminergic
    disorders
  affecting the highly educated and
      successful, 151
  and bipolar disorder, 82, 85
  and breakthrough pharmacological
      treatments, 161
  dopamine involvement, 16
  emotion, 65
  epigenetic influences, 102
  excessive dopamine activity, 75
  financial cost of, 159
  heritability, 4
  obsessive-compulsive disorder, 87
  prevalence, 76
  rarity outside industrialized nations,
      151
  schizophrenia, 82, 91
  Tourette's syndrome, 81–2
autonomic nervous system, and dopamine,
    33–5

"Big Bang" (cultural explosion in *Homo
    sapiens*), 101, 104, 114–15, 121–2,
    153–4
bipedalism
  and brain lateralization, 18
  and heat absorption, 104
  and thermoregulation, 108–9
bipolar disorder, 75, 76
  and mania, 84–6
  obsessive-compulsive disorder, 87
birds, dopamine in, 15
Blombos Cave artifacts, 115
body dysmorphic disorders, 87
Book of Prophecies (Columbus), 137
boredom, risk of, 163, 171

brain, anatomy, 23–5
brainstem, 24
Broca's area, 58

caudate, 27
chase (persistence) hunting, 109–10
Cheney, Dick, 161
chimpanzees
    ability to engage in mirror self-
        recognition, 2
    genetic coding differences to humans,
        4–5
    heart position, 8
cholinergic activation, and dreams, 51
cholinergic inhibition, and hallucinations,
    51
chorion, 76–7
Churchill, Winston, 132
climate change, 158, 159–60, 167
clozapine, 94
cocaine, 21, 23, 28
    maternal ingestion of, 102–3
cognitive decline due to aging, 54–5
cognitive flexibility (mental shifting)
    and intelligence, 59–61
    dopamine involvement, 16, 54
Columbus, Christopher, 130, 136–9,
    148–9
computers, human brain compared to, 13
concordance, 76
constructional apraxia, 53
context-independent cognition, 2
cortex, 23
cortical layers, 31
cosmic experiences, 92–3
Crab-eating Macaque monkeys, 118
cradle of humanity, hominids
    environmental adaptations in, 104–8
creativity, dopamine involvement, 16
cultural exchange, 119–21
cyclothymia, 84–5

D1 receptor
    importance to working memory, 21
    neuroanatomy, 28
D2 receptor
    importance to clinical disorders, 21
    neuroanatomy, 28
delusions of grandeur, 35, 70–1
depression, financial cost of, 159
depth perception, 93
dichorionic twins, 76–7
diencephalon, 24–5
distant space and time
    and dopamine, 38–41

attention to distant cues, 41–5
and intelligence, 57
dizygotic/fraternal twins vs. monozygotic/
    identical twins
    difference in prenatal environment, 4,
        103
    hand dominance, 8
dolphins, 1
dopamine
    and advanced intelligence, 13–17
    and autonomic nervous system, 33–5
    and intelligence, 53–7
    and left hemisphere, 31–3
    and thermoregulation, 110–11
    chemical structure, 19–20
    expansion during human evolution, 15,
        17–18
    lateralization of, 14
    neural transmission process, 20–3
    neuroanatomy, 23–31
    number of studies into, 16–17
dopamine beta-hydroxylase, 77
    deletion of gene in mothers, 77, 83
dopamine receptors, 21
dopamine-mediated hyper-religiosity
    mania, 85
    obsessive-compulsive disorder, 87
    schizophrenia, 85, 93
dopaminergic imperative
    as relinquishable, 172
    definition, 155
dopaminergic mind, see also
        protodopaminergic mind
    definitions, 3
    evolution of, 101, 114–21
    limits of, 161
    measures necessary to alter, 170–2
    tempering with individual behavior,
        161–5
    tempering with societal change, 165–70
dopaminergic personality, 29, 66–7, 73–4
    in human history, 130–2, 147–9,
        153–4
dopaminergic society
    modern hyperdopaminergic, 149–53
    transition to, 123–9
dorsal-ventral axis, 23
dreams/dreaming, 30, see also extrapersonal
        experiences
    similarity with hallucinations, 50
dressing apraxia, 53

ecological footprint, 168
"ego" analogy, 67
Einstein, Albert, 131, 144–7, 148, 149

emotion, and dopamine, 64–6
"end of science", 157
Enlightenment, 168
environmental impacts, climate change, 158, 159–60, 167
epigenetic inheritance, 4
epigenetic transmission, importance of, 101–4
estrogen, 72
evolution, *see* human evolution
executive intelligence, 16, 54, 55
exploratory behavior, 42–5, 166–7
extrapersonal experiences, 49–53
eyes, *see* saccadic eye movements; third eye; upward eye movements

famous persons, 130–2
Faraday, Michael, 91
feeding, 31, 41, 47
Feynman, Richard, 131
Flandrian Transgression, 127
focal-extrapersonal brain system, and lateral dopaminergic personality, 39
Foxp2 gene mutation, 6–7
frontal lobe, 24
future orientation, 150
Future Shock (Toffler), 151

GABA (gamma-aminobutyric acid), 14
Gaia, 166
gambling, 52
gamma-aminobutyric acid (GABA), 14
Gandhi, Indira, 166
generative grammar, 63
generativity/creativity, 2
    and intelligence, 63–4
genetic coding, humans vs. chimpanzees, 4–5
genetic inheritance, of disorders, 77
genius and madness, 69, 75–6
Garden of Eden, 128
global warming, 158, 159–60, 167
glorification of military conquest, 168
glutamate, 14, 20
goal-directedness, 46–9
gracile australopithecines, 104–5
grandeur, delusions of, 35, 70–1
growth hormone, 111

hair growth reduction, evolution of, 112
hallucinations, *see also* extrapersonal experiences
    schizophrenia, 93–4
    similarity with dreams, 50

haloperidol, 21, 23
    maternal ingestion of, 102–3
handedness in humans
    evolutionary advantage of, 8
    right-hand dominance, 7
    twins, 8
heat/thermal stress, 18, 95, *see also* bipedalism, and heat absorption
hedonic activity, 48
hierarchical religious institutions, 127, 169–70
hikikomori, 169
hippocampus, 24, 27
hoarding, 41–2, 126
Homer, *Odyssey*, 129
hominins, 104
Homo erectus, 105–6, 113
Homo ergaster, 105–6
Homo habilis, 1, 105, 113
Homo heidelbergensis, 106
Homo helmei, 106
Homo sapiens, 113
Homo sapiens neanderthalensis (Neanderthals), 106, 107–8, 117–18
Homo sapiens sapiens, 106–7
Hughes, Howard, 132
human cognition, distinctive attributes of, 2–3
Human Development Index, 167–8
human evolution
    advantages of handedness, 8
    dopamine expansion, 15, 17–18
    dopaminergic mind, 101, 114–21
    hair growth reduction, 112
    intelligence, 121–2
    protodopaminergic mind, 104–14
    role of population pressures and cultural exchange, 119–21
    shellfish consumption, 18, 117–19
human intelligence
    as genetically selected, 3–10
    evolution of, 121–2
hunter-gatherer societies, 124–5
Huntington's chorea, 84
Huntington's disease, 83–4, *see also* hyperdopaminergic disorders
    dopamine involvement, 16
    excessive dopamine activity, 75
    generativity/creativity, 63–4
    schizophrenia, 91
hyperdopaminergic disorders, 97–100, *see also* attention-deficit/hyperactivity disorder; autism; Huntington's disease; obsessive-compulsive disorder; Parkinson's disease;

phenylketonuria; schizophrenia;
    Tourette's syndrome
  and breakthrough pharmacological
    treatments, 161
hyperdopaminergic society, 149–53
  advent of, since World War II, 154
  mental health of, 158–9
  pillars of, 165–70
"hyperdopaminergic" syndrome, 75–9
hyperthyroid patients, psychosis, 92
hypnogogic hallucinations, 50
hypnopompic hallucinations, 50
hypomania, 75
hypothalalamus, 24
hypothalamus, 24–5
hypoxia, 77

"id" analogy, 67
impulse-control disorders, 87
incentive motivation, 46–7
intelligence, see also executive intelligence;
    human intelligence
  and abstraction, 61–2
  and cognitive flexibility, 59–61
  and distant space and time, 57
  and dopamine, 53–7
  and generativity/creativity, 63–4
  and motor programming and
    sequencing, 57–9
  and working memory, 59
intelligence tests, 55–6
internal locus of control, 35, 70
interplanetary space exploration, 166–7
iodine-deficiency syndrome (cretinism),
    16, 54
island fever, 151

Jaynes, 128, 129
Judeo-Christian religions, 170
Julius Caesar, 132

Kafka, Franz, 91
Khoisan people, 115
King, Martin Luther, Jr, 48
Klasies River excavations
  shellfish consumption, 118
  skulls, 106
!Kung San, 125

language, see also speech, sequencing of
  and abstraction, 61
  and cognitive flexibility, 60
  involving other than speech
    and hearing, 9
  linguistic competence, 56–7

reliance on working memory, 59
language gene, requirements of, 9
latent inhibition, 60–1, 64
lateral dopaminergic pathway (nigrostriatal
    pathway), 26–7, 29
lateral dopaminergic personality, 29
  and focal-extrapersonal brain system, 39
lateral prefrontal cortex, personality
    traits, 67
lateral regions, 23
lateral-dopaminergic traits, 69–71
lateralization of the brain
  abstract/analytical/propositional
    thought, 62
  and bipedalism, 18
  left/right hemisphere comparison,
    13–14
  not accounted for by genetic factors, 4,
    7–9
L-dopa, 84, 86, 89–90, 94
left hemisphere
  and dopamine, 31–3
  generativity/creativity, 64
left-hemispheric (masculine) style, 71–3
left-hemispheric bias, in working
    memory, 59
Levant region, 107
limbic cortex, 24
linguistic competence, 56–7
locus-of-control, 52, 70, 79, 150

mania
  and bipolar disorder, 84–6
  and schizophrenia, 85, 91, 92
  dopamine involvement, 16
  excessive dopamine activity, 75
  generativity/creativity, 64
  ventromedial dopaminergic traits, 68
masculine dominance in society, 165–6
maternal fever, 77
maternal phenylketonuria, 90
Mathematical Principles of Natural
    Philosophy (Newton), 140
meat consumption, 105, 112
medial frontal cortex, 28
medial regions, 23
Meir, Golda, 166
mental health of future society, 158–9
mental time travel, 2–3, see also "off-line"
    thinking
Messiah complex, 92
methylphenidate, 23, 80
mid-sagittal view, 23
migration, human, 105, 113, 115, 116,
    117, 119, 166, 171

military conquest, glorification of, 168
Minoan civilization, 127
monoamine oxidase, 21
monochorionic twins, 76–7
Mother Earth, 166
Mother Nature, 166
motor behavior, dopamine stimulation,
    37–8
motor control, dopamine involvement, 16
motor planning, dopamine involvement, 16
motor programming and sequencing, and
    intelligence, 57–9
Mustafa Kemal Atatürk, 132
myopia for the future, 45

Naples High Excitability rat, 80
Napoleon Bonaparte, 142–4, 147, 148,
    149
Nash, John, 69, 91
Neanderthals (*Homo sapiens
    neanderthalensis*), 106, 107–8, 117–18
neural transmission process, dopamine,
    20–3
neurotransmitters
    lateralization of, 14
    neurochemistry, 20
Newton, Sir Isaac, 91, 139–42, 148–9
noradrenaline, *see* norepinephrine
norepinephrine (noradrenaline)
    and intelligence, 56
    and neural transmission of dopamine, 21
    in arousal systems, 39
    in dorsal parietal-occipital areas of the
        brain, 30
    lateralization of, 14
    neurochemistry, 19
    production cells, 25
norepinephrine, association with
    interoceptive senses, 40
novelty-seeking, 44
nucleus accumbens, 27, 68

obsessive-compulsive disorder, 28, 86–8,
    see also hyperdopaminergic disorders
    and bipolar disorder, 85
    and other hyperdopaminergic disorders,
        99–100
    dopamine involvement, 16
    excessive dopamine activity, 75
    prevalence, 76
    schizophrenia, 87, 91
    ventromedial dopaminergic traits, 68
"obsessive-compulsive/autistic" syndrome,
    82
occipital lobe, 24

oculogyric crises, 93
Odyssey (Homer), 129
"off-line" thinking, 2–3, 30, 61, 122, 123
Olduvai Theory, 160
olfactory system, 5
    connection to medial dopaminergic
        motivation systems, 40
Omo Kibish skulls, 106–7
Omo valley, 105
orangutans, 2
orbitofrontal frontal cortex, 28
original affluent society, 124
orofacial movements, 57
otolith organs, 33
out-of-body experiences, 49

PAH gene, 90
parasympathetic system, 33, 34
parietal lobe, 24
Parkinson's disease, 88–90, 111, *see also*
    hyperdopaminergic disorders
    dopamine involvement, 16
    dopamine treatment and pathological
        gambling, 88
    generativity/creativity, 63–4
    nigrostriatal dopaminergic system, 27
    Raven's Progressive Matrices test, 62
    temporal analysis/processing speed, 63
    typically first appears on the left side of
        the body, 32
    upper-field behavioral bias, 42
parrots, 1–2
peripersonal brain system, 38
persistence (chase) hunting, 109–10
phenylalanine, 23
phenylalanine hydroxylase, 90
phenylketonuria, 16, 54, 63–4, 90–1, 111,
    see also hyperdopaminergic disorders
    epigenetic influences, 102
phenylpyruvic acid, 90
Pinnacle Point, evidence for shellfish
    consumption, 117
piribedil, 63
post-dopaminergic consciousness, 171
Pound, Ezra, 91
prefrontal cortex, dependence on
    dopamine, 15–16
prefrontal leucotomies, 29
prenatal iodine-deficiency, epigenetic
    transmission of, 102
Principia (Newton), 140
Protestant work ethic, 152
proto-civilizations, 126–9
protodopaminergic mind, 101
    evolution of, 104–14

pursuit of scientific discovery, 168–9
pursuit of wealth, 167–8
putamen, 27

Raven's Progressive Matrices Test, 55–6,
    61–2
religious experiences, *see also* dopamine-
    mediated hyper-religiosity
  activation of ventral brain regions, 51–2
  and dreaming and hallucinations, 50
  upward bias, 51
religious systems
  agrarian societies, 126
  hierarchical/male dominated, 127,
    169–70
  hunter-gatherer societies, 125
  Renaissance, 168
reward prediction, 45, 46–7
Rhino Cave findings, 115
risperidone, 82
robust australopithecines, 104–5
Rolandic fissure, 24

saccadic eye movements, 38, 39, 40, 44–5,
    see also anti-saccade
Saharasia, 126–7
Sahlin, 124
scavenging, 109, 110, 112, 113
schizophrenia, 91–5, 111, *see also*
    hyperdopaminergic disorders
  and bipolar disorder, 85–6
  and dopamine-mediated hyper-
    religiosity, 85, 93
  and mania, 85, 91, 92
  dopamine involvement, 16
  emotion, 65
  excessive dopamine activity, 75
  generativity/creativity, 64
  haloperidol, 23
  mesolimbic system, 28
  obsessive-compulsive disorder, 87, 91
  prevalence, 76
  rarity outside industrialized
    nations, 151
  upward eye movements, 42
  ventromedial dopaminergic traits, 68
schizotypy, 92
scientific discovery, pursuit of, 168–9
seasonal affective disorder, 85
sentences, responding to, 6
serotonin (5-hydroytryptmine), 14, 19–20,
    21, 25, 56
  inhibitory action on dopaminergic
    systems, 78
sexual behavior, 28, 34, 38

female, 41, 47–8
male, 40, 47, 48
shellfish consumption, importance in
    human evolution, 18, 117–19
social grooming, 41
space exploration, 166–7
species extinctions, 156, 158
speech, sequencing of, 57–8
"stargazer" rat, 37, 42
stereotypy, 37
stress, *see also* thermal/heat stress
  beneficial rise of dopamine during, 36,
    70, 78–9
  controllable versus uncontrollable
    stressors, 69
  maintaining motivation and attentional
    focus, 34–5
  mesolimbic system, 28, 29
Strindberg, August, 91
subcortical forebrain, 25
subcortical regions, 23
substance abuse, 68, 75
  and bipolar disorder, 85
superstitious behavior, 52
sweating, 109
Sydenham's chorea, 84
Sylvian fissure, 24
sympathetic system, 33–4
synapses, 20–1, 22

technology, over-zealous faith in, 168–9
tegmentum, 25
telic personality, 70
temporal analysis/processing speed,
    62–3
  dopamine involvement, 16
temporal lobe, 24
testosterone, 34, 47, 72, 131
thalamus, 24–5
thalidomide, 77, 82
Thatcher, Margaret, 166
theory of mind, 2, 65–6
thermal/heat stress, 18, 95, *see also*
    bipedalism, and heat absorption
thermoregulation, 108–14
third eye, 51
thyroid gland, and shellfish and dopamine,
    117
thyroid hormones, 17
Toffler, Alvin, *Future Shock*, 151
tool-making, 114
Tourette's syndrome, 95, *see also*
    hyperdopaminergic disorders
  and bipolar disorder, 85
  autism, 81–2

Tourette's syndrome (*cont.*)
  dopamine involvement, 16
  excessive dopamine activity, 75
  obsessive-compulsive disorder, 87
  prevalence, 76
trichotillomania, 87
twins, *see* dizygotic/fraternal twins vs.
    monozygotic/identical twins;
    monochorionic twins
tyrosine, 23, 90
  maternal ingestion of, 102–3

upper-field behavioral biases, 40, 42
upward eye movements
  during extrapersonal experiences, 51
  during higher mental activity, 42, 58
  in schizophrenia, 42
  oculogyric crises, 93
US society, 152–3

valproic acid, 82
ventromedial dopaminergic
    personality, 29
  and action-extrapersonal brain system,
    39–40

ventromedial dopaminergic traits,
    67, 68–9
ventromedial/medial dopamine system
    (mesolimbic pathway), 25–6, 27–9, 48
vestibular assymetry
video-game playing, 88, *see also*
    hikikomori

wanting versus liking, 48, 68
war
  development of, 154, *see also* glorification
    of military conquest
wealth, pursuit of, 167–8
wildebeest, 110
winter births, excess of
  bipolar disorder, 86
  famous persons, 130
Wisconsin Card-Sorting Test, 60
working memory
  and intelligence, 59
  dopamine involvement, 16
  importance of, 56
  importance of D1 receptor, 21

yin and yang, 72